FLOW OF FLUIDS
through
Porous Materials

FLOW OF FLUIDS
through
Porous Materials

ROYAL EUGENE COLLINS, Ph.D.

*Professor of Physics
at the University of Houston and
Consultant to the Humble Oil
and Refining Company*

THE PETROLEUM PUBLISHING COMPANY
TULSA, 1976

PREFACE TO THE REPRINT

Although this text has been available in Russian (Moscow University 1965) and more recently in Japanese (Kyoto University 1975), it has been out of print in English for more than a decade, therefore I am most pleased that Petroleum Publishing Company has elected to make it available again in English.

In view of the growing demand for improved technology in secondary and tertiary recovery of petroleum, a great need exists for a simple text on fundamentals of flow through porous materials to assist in training engineers and scientists, and although many developments in the field have occurred in the fifteen years since original publication of this volume it still seems to meet this need.

R. E. COLLINS

Houston, Texas
February, 1976

CONTENTS

1. STRUCTURE AND PROPERTIES OF POROUS MATERIALS

1.10: Structure and Classification

In the most general sense, a porous material is a solid containing holes. However, a hollow metal cylinder, for example, is not usually classed as a porous material; consequently, a more precise specification of the term *porous material* is required. For the purposes of this study, a solid containing holes or voids, either connected or non-connected, dispersed within it in either a regular or random manner will be classed as a porous material provided that such holes occur relatively frequently within the solid.

A great variety of natural and artificial materials are porous; a bucket of sand, a piece of limestone, a tuft of cotton or a loaf of bread are examples of porous materials. In Figure 1-1 some examples of porous materials are shown. From these, it is obvious that great variations in the size and structure of pores exist in such materials. Some classification of pores is possible, however.

Pores are either *interconnected* or *non-interconnected*. A fluid can flow through a porous material only if at least some of the pores are interconnected. The interconnected pore space is termed the *effective* pore space, while the whole of the pore space is termed the *total* pore space.

The voids within a porous material can further be classified according to their size. Three main classifications are possible, based on the behavior of fluids within the void space. In the smallest void spaces molecular forces between the molecules of the solid and those of the fluid are significant. These tiniest void spaces are termed molecular *interstices*. In the largest void spaces the motion of a fluid is only partially determined by the walls of the void: these largest spaces are referred to as *caverns*. Those spaces which are intermediate in size between molecular interstices and caverns are termed *pores*. A further classification of these pores is sometimes made, particularly in limestone or dolomite rocks. Solution cavities of small size are referred to as *vugs* and the void space formed by these is called *vugular* pore space.

An additional classification of porous materials divides them into two groups, *ordered* or *random*. The meaning is obvious. A regular packing of uniform spheres is ordered while a loaf of bread is random in its porous structure.

1

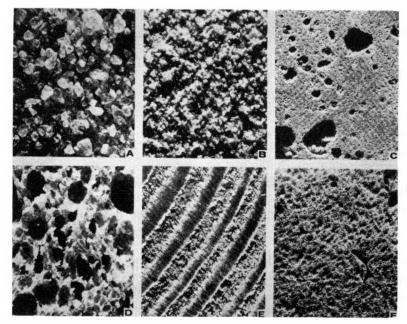

Figure 1–1. Examples of natural porous materials (×10): (a) beach sand, (b) sand-stone, (c) limestone, (d) rye bread, (e) wood, (f) human lung.

1.20: Structure and Properties

Most naturally and artificially porous materials have a random void structure. In fact the structure of such materials can be described only in statistical terms. Even so, it is possible to treat the flow of fluids through such materials on a macroscopic basis in precise terms. The situation is much like that in the kinetic theory of gases; on a microscopic scale the variables in question must, because of their great number and complexity, be treated as random variables but on the macroscopic scale the system can be treated in terms of a few completely determinable quantities.

Many theories have been devised which attempt to relate in a detailed manner the macroscopic properties of porous materials to the statistical properties of their microscopic structure. Most of these theories attempt to relate "pore size distribution" to the macroscopic properties of the material.[3] Some attempts have also been made to relate the "grain size distribution"[12] of unconsolidated materials to their macroscopic properties. While such theories contribute greatly to our understanding of basic physical processes within porous media, they do not, in general, contribute to the solution of problems on a macroscopic scale.

The macroscopic theory of fluid flow through porous materials can be developed in either of two ways. One can begin with a statistical microscopic theory and show how this leads to certain macroscopic laws, just as the kinetic theory of gases predicts the macroscopic Boyle's Law, or one can begin with the macroscopic laws as empirical laws established by experiment.

Since none of the existing statistical theories of flow through porous media satisfactorily accounts for all macroscopic phenomena, the latter course is followed in this volume. Where consideration of microscopic processes is necessary to the understanding of macroscopic phenomena the details of the structure of porous materials are brought into the discussion.

The macroscopic properties of porous materials which are important in the study of fluid flow through porous media are defined and discussed in the following sections. All of these properties are bulk properties and as such have significance only for samples of porous materials containing relatively large numbers of pores.

1.30: Porosity

The porosity of a porous material is the fraction of the bulk volume of the material occupied by voids. The symbol usually employed for this parameter is ϕ though f is sometimes used. Thus

$$\phi = \frac{V_P}{V_B} = \frac{\text{Volume of pores}}{\text{Bulk volume}} \tag{1-1}$$

which is a dimensionless quantity.

Since that portion of the bulk volume not occupied by pores is occupied by the solid grains or matrix of the material, it follows that

$$1 - \phi = \frac{V_s}{V_B} = \frac{\text{Volume of solids}}{\text{Bulk volume}} \tag{1-2}$$

Two classes of porosity can be defined, namely, *absolute* or *total*, and *effective* porosity. Absolute porosity is the fractional void space with respect to bulk volume regardless of pore connections. Effective porosity is that fraction of the bulk volume constituted by interconnecting pores. Many naturally occurring rocks, such as lava and other igneous rocks, have a high total porosity but essentially no effective porosity.

Effective porosity is an indication of permeability but not a measure of it.

1.31: Porosity and Structure

The manner in which porosity depends on the structure of a porous material can be seen by considering simple models, such as regular packings

Cubic   Rhombohedral

Figure 1–2. Packing of uniform spheres. (*After Graton and Fraser, 1935.*)

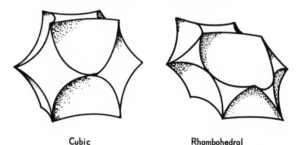

Cubic   Rhombohedral

Figure 1–3. Pore space in packing of uniform spheres. (*After Graton and Fraser, 1935.*)

of uniform spheres or rods. For such systems, one can consider unit cells such as shown in Figure 1-2.

Figure 1-2 shows two types of regular packing for uniform spheres. Case 1 is the cubic packing and Case 2 is the rhombohedral packing. These two arrangements represent the "loosest" and "tightest" packing that can be obtained with uniform spheres. The shapes of the pores contained within the unit cells for these two cases are shown in Figure 1-3.

These "pores" are extremely simple in form when contrasted to those of a natural material, as is shown in Figure 1-4. This figure shows the actual form of the pore structure in a consolidated sandstone. This was obtained by impregnating the sandstone with Woods metal and then dissolving the sand.

Graton and Fraser[8] consider six different packings of uniform spheres. The porosities of these various packings fall between the limiting values of 0.2595 for the rhombohedral packing and 0.4764 for the cubic packing.

It is nearly impossible to obtain anything approaching a regular packing by pouring a container full of spheres. "Bridging" invariably occurs and very high porosities result. Sand packs of uniform grain size usually consist of small regions of more or less regular packing separated by regions of

Figure 1–4. Cast of pore space in sandstone.

irregular packing in which "bridging" has occurred. In the bridged regions, the porosity is invariably greater than that corresponding to the "loosest" regular packing.

Theoretically, the porosity of a packing of uniform spheres should be independent of the size of the spheres, but for natural materials this proves not to be the case. Actual measurements show that, for sands of essentially uniform grain size, the porosity increases as the grain size decreases. As a general rule, the smaller the grains the greater will be the porosity in any naturally unconsolidated material of uniform grain size.

For naturally occurring unconsolidated materials of non-uniform grain size, the porosity is dependent on the distribution of grain size. A variety of grain sizes permits the smaller grains to fill the pores formed by larger grains thus resulting in lower porosities. Generally, a poorly sorted material will have a considerably lower porosity than will a fine-grained but well-sorted material.

1.32: Effects of Consolidation and Compaction on Porosity

Compaction is the process of volume reduction due to an externally applied pressure. Consolidation refers to the binding together of the elements of the solid matrix by a cementing material.

For some naturally porous materials, such as clays and silts, and fibrous materials, such as cloth and paper, compaction may produce significant changes in porosity. Extremely hard materials like silica sand suffer only slight changes in porosity under rather large compaction pressures.

For extreme compaction pressures, all materials show some irreversible change in porosity. This is due to distortion and crushing of the grains or matrix elements of the material and, in some cases, recrystallization.

The most significant factor in the determination of the porosities of rocks is cementation. Consolidated sedimentary rocks are regarded as initially unconsolidated sands which have undergone significant cementation during geologic time. This cementation, which exists principally at what were originally grain contacts, can be distinguished, in most cases, from the original grain material by its chemical composition. As the pore space is filled with cementing material, great reduction in porosity can occur.

Another type of cementing occurs in some artificially consolidated porous materials. Ceramics, sintered glass and sintered metals represent porous materials which have been consolidated essentially by fusion.

Naturally occuring clays exhibit the greatest range of porosity of all natural materials. Clays occur in the form of plate-like grains. These materials are generally speaking very hygroscopic; some, such as montmorillonite, can absorb several times their own volume in water, undergoing considerable swelling in the process.

Since clays are composed of very small plate-like grains, it should be expected that they would be very susceptible to compaction. Consequently, the porosities of clays should be lower for increasing depth below the earth's surface. Athy[1] has, indeed, found that the variation of clay porosity with depth can be represented by

$$\phi = \phi_0 e^{-\alpha z} \tag{1-3}$$

where ϕ_0 is the average porosity of surface clays, α is a constant and z is depth below the surface. His data show that while the porosity of surface clay is between 0.40 and 0.50, the porosity of shale (which is highly compacted clay) at a depth of 6,000 feet is only 0.05.

Although cementation and compaction tend to reduce the porosities of natural rocks, other secondary factors tend to increase their porosities. Chemical leaching and physical erosion due to the flow of ground water through the porous rocks enlarge the pores. Extreme examples of this process are seen in natural caverns, such as Carlsbad Cavern in New Mexico.

1.33: Measurement of Porosity

From the definition of porosity, it is evident that the porosity of a sample of porous material can be determined by measuring any two of the three quantities: bulk volume, pore volume or solids volume.

Direct Method. The most direct procedure is to measure the bulk volume, crush the specimen, remove all pores, and then measure the remaining volume of solids. This technique is often used for brick and ceramics. This method yields a measure of total porosity.

Gas Expansion Method. Perhaps the most widely used method of measuring effective porosity is that based on gas expansion. By enclosing the specimen, of known bulk volume, in a container of known volume under a known air (or gas) pressure and then connecting this with an evacuated container of known volume, the pore volume can be computed from the observed pressure change using the Boyle-Mariotte gas law. Thus

$$\text{Pore Volume} = V_B - V_a - V_b \frac{P_2}{P_2 - P_1} \qquad (1\text{-}4)$$

where

V_B = bulk volume of sample
V_a = volume of sample chamber
V_b = volume of second (evacuated) chamber
P_1 = initial pressure, and
P_2 = final pressure.

A variety of instruments of many different forms has been developed using this basic principle. Methods based on gas expansion are not so accurate as another technique.

Mercury Injection Method. The mercury injection method of measuring effective porosity is based on the fact that, due to the surface tension and non-wetting properties of mercury, a porous sample can be immersed in mercury without entry of mercury into the sample at atmospheric pressure. Thus, the bulk volume of the sample can be determined by displacement of mercury from a sample chamber of known volume.

If the sample chamber is closed and the hydrostatic pressure of mercury in the chamber is increased to a very great value, the mercury will enter the pores compressing the trapped air in the pores to negligible volume. The volume of mercury injected is, therefore, equal to the pore volume.

An advantage of this method is that both bulk volume and pore volume are directly determined. This method is not very precise since the volume occupied by compressed air is not determined. It is undesirable since the sample invariably contains mercury contamination even after extensive cleaning procedures and hence is not suitable for further tests.

Density Methods. Since the mass of a porous material resides entirely in the grains, or matrix, it follows that

$$M = \rho_s V_s = \rho_B V_B \tag{1-5}$$

where

$$M = \text{mass of sample}$$
$$\rho_s = \text{density of grain material, and}$$
$$\rho_B = \text{bulk density of sample.}$$

In view of equation (1-2), it follows that

$$\phi = 1 - \frac{\rho_B}{\rho_s} \tag{1-6}$$

Bulk density can readily be determined by weighing the sample and measuring the bulk volume by a volumetric displacement technique. Mercury immersion can be used but it is best to apply a waterproof coating and employ water immersion.

The density of the solid material can be determined by crushing a sample of the material, weighing and then employing a displacement technique on the parts for volume determination. This method yields total porosity.

Imbibition Method. A very direct method of measuring effective pore volume, and hence effective porosity, is widely employed in the petroleum industry. Since most clean rocks are strongly water wet, they imbibe water readily. Thus, if a sample of rock is immersed in water under vacuum for a week or so, the pore space becomes completely water filled. Then, the mass of the saturated sample is

$$M' = M + \rho_w V_P \tag{1-7}$$

where ρ_w is the density of water (≈ 1) and M is the dry mass of the sample. Hence

$$V_P = \frac{M' - M}{\rho_w} \tag{1-8}$$

With the sample completely saturated with water, a volumetric displacement measurement in water gives directly the value of V_B without any coating procedure and ϕ can be computed.

Except for the great length of time required for complete saturation to occur, this is perhaps one of the best methods of porosity measurement in current use.

Statistical Method. Since, for naturally occurring porous materials, the porous structure is of a spatially random nature, the plane porosity of a random section must be the same as the volumetric porosity. The plane porosity is defined as the fraction of the area of a plane section constituted by voids or pores. The probability for a random point on such a section to lie within a pore is simply the porosity, ϕ.

This principle has been employed by Chalkley, Cornfield and Park[5] to measure porosity. A pin is dropped many times in a random manner on an enlarged photomicrograph of a section of the porous material. It can be shown that in the limit, as the number of pin tosses is increased, the ratio of the number of times the point falls in a pore to the total number of tosses approaches the value ϕ.

Note that, since both isolated and connected pores are exposed, this method yields total porosity.

1.40: Specific Surface

The *specific surface*, Σ, of a porous material is defined as the interstitial surface area of the pores per unit of bulk volume of porous material. It is obvious that finely structured materials will exhibit a much greater specific surface area than will coarse materials.

Specific surface plays an important role in the design of filter columns, reactor columns and ion exchange columns. It is also an important parameter with regard to the fluid conductivity or permeability of a porous material. This point is discussed at greater length in connection with the Kozeny equation. (Section 1.51.)

Since specific surface is the ratio of area to volume its dimensions are L^{-1}.

1.41: Measurement of Specific Surface

Since the internal surface of any natural porous material is of extreme complexity the specific surface area, Σ, can only be determined by statistical or indirect means. Three such methods are described here.

Statistical Method. The statistical method of determining porosity developed by Chalkley, Cornfield and Park[5] was also extended by them to the determination of specific surface. In their method an enlarged photomicrograph of a section of the porous material is used.

If a needle of length l is dropped a great number of times on the picture and counts are made of the number of times the end points fall within pores, and the number of times the needle intersects the perimeter of pores, then an equation based on probability theory can be employed to compute Σ.

Denoting the number of times the points fall within pores by h and the number of perimeter intersections by c, this equation is

$$\Sigma = \frac{4\phi c}{lh} \, m \qquad (1\text{-}9)$$

where m is the over-all magnification in the picture.

This method is the best method in current use for determining specific surface.

Adsorption Methods. The quantity of a vapor which can be adsorbed on a surface is dependent on the area of the surface. Several theories[2] have been employed to determine surface area in this way. However, all adsorption methods of determining specific surface are subject to the same criticism. The quantity of a gas or vapor adsorbed is proportional to a surface area which includes the tiny molecular interstices of the porous material, whereas the surface area pertinent to fluid flow does not include this portion of surface area.

Methods Based on Fluid Flow. The Kozeny equation, or the Kozeny-Carman equation, both of which are discussed in section 1.51, relates the fluid conductivity, or permeability, of a porous medium to the specific surface. Consequently, measurements of fluid conductivity have been used extensively to compute specific surface from these equations.

Since, as is pointed out in section 1.51, the Kozeny equation is not strictly correct, the values of specific surface determined in this manner are subject to serious uncertainties.

The determination of specific surface by use of the Kozeny equation is compared to other methods by Brooks and Purcell.[2]

1.50: Permeability

Permeability is that property of a porous material which characterizes the ease with which a fluid may be made to flow through the material by an applied pressure gradient. Permeability is the *fluid conductivity* of the porous material.

That a parameter characterizing the fluid conductivity of a porous material can be meaningfully defined was first demonstrated by Darcy in 1856.[6] In fact, the equation which defines permeability in terms of measurable quantities is called *Darcy's law.**

If horizontal linear flow of an incompressible fluid is established through a sample of porous material of length L in the direction of flow, and cross-sectional area A, then the permeability, K, of the material is defined as

$$K = \frac{q\mu}{A(\Delta P/L)} \qquad (1\text{-}10)$$

---

* See sections 3.11 and 3.30, chapter 3.

Here q is the fluid flow rate in volume per unit time, μ is the viscosity of the fluid and ΔP is the applied pressure difference across the length of the specimen.

The value of the permeability, K, is determined by the structure of the porous material. From the defining equation (1-10), it is seen that K has dimensions of length squared. K is roughly a measure of the mean square pore diameter in the material. Many porous materials have a directional quality in their structure. As a consequence, the permeabilities measured with flow perpendicular to each face of a cube of such material are not all equal. Such materials are termed anisotropic porous media. The properties of such materials are discussed further in section 3.32.

The unit most widely employed for permeability is the darcy (d). This unit is defined: For a material of one darcy permeability a pressure difference of 1 atmosphere will produce a flow rate of 1 cubic centimeter per second of a fluid with 1 centipoise viscosity through a cube having sides 1 centimeter in length. Thus

$$1 \text{ darcy} = \frac{1(\text{cm}^3/\text{sec}) \cdot 1(\text{cp})}{1(\text{cm}^2) \cdot 1(\text{atm/cm})} \tag{1-11}$$

For very "tight" materials the millidarcy, (md) = 0.001 d, is used.

Discussion of methods of measuring permeability will be deferred until a more complete discussion of this fundamental law of flow is given.

1.51: Structural Interpretation of Permeability

The permeability of a porous material as defined by Darcy's law is a macroscopic property of the material. As such, it has significance only for samples sufficiently large to contain many pores.

It is obvious that permeability must be determined by the geometry of the porous structure in a more or less statistical fashion. Many attempts have been made to construct a theory which relates this structure to permeability. A rather complete review of such theories is given by Scheidegger.[19] A few of these theories are described here.

The theory of Kozeny[11] treats the porous medium as a bundle of capillary tubes of equal length. These tubes are not necessarily of circular cross section. By considering the solution of the classical hydrodynamic equations for slow, steady flow through such a system, Kozeny was able to show that the permeability for such a system must have the form

$$K = \frac{c\phi^3}{\Sigma^2} \tag{1-12}$$

where c is a dimensionless constant depending only on the geometrical form of the capillary tube cross section and K is in units of length squared. For a circle: $c = 0.50$; for a square, $c = 0.5619$; and for an equilateral triangle, $c = 0.5974$. The value c is called the Kozeny constant.

Numerous modifications of the Kozeny equation have been proposed. One such modification is proposed to account for the fact that the tubes of flow in a porous medium are not straight, and hence the path length of flow is greater than the length of the sample of porous material. Thus, if *tortuosity*, τ, is defined as the ratio of flow-path length to sample-path length, the modified Kozeny equation is

$$K = \frac{c\phi^3}{\tau\Sigma^2} \tag{1-13}$$

Other modifications are discussed by Brooks and Purcell.[2]

It is difficult to verify equations of the Kozeny type since the factors Σ and τ are difficult to determine independently. Furthermore, c is in practice only an empirical factor which shows considerable variation from sample to sample. Even so, the Kozeny theory does show quite conclusively that ϕ, Σ and K are interrelated.

Another approach to the relation between pore structure and permeability is also based on a capillary-tube model. This is the calculation of permeability from "pore-size distribution." In this scheme, the porous material is treated as a bundle of capillary tubes having equal lengths and circular cross sections with a distribution of radii. Since the flow through each tube is given by the Hagen-Poiseuille law, the flow through the system can be related to the radius-distribution function, which, in turn, yields an expression for permeability. Burdine, Gournay and Reichertz[3] have applied this theory with pore-size distributions determined by the mercury-injection method (section 2.30) rather successfully to sedimentary rocks.

In the case of unconsolidated porous materials, such as sand, one would expect the geometry of the pore space to be very closely related to the shape and size distribution of the grains of solid. Krumbien and Monk[12] have shown that it is possible to relate certain parameters of the grain-size distribution function of sands to their permeability.

For sands having moderately spherical grains and essentially a normal (Gaussian) distribution of grain size (as determined by a sieve analysis), these investigators found a correlation between permeability, the geometric mean particle diameter and the standard deviation of the distribution function.

Without going further into discussion of the various structural theories of permeability, certain general conclusions are possible. Permeability must be proportional to some sort of mean square pore diameter, or radius squared, and the spread of pore size must also be an important factor in the determination of the permeability. Of course, these factors also determine the specific surface of the material and hence the Kozeny theory also relates to pore size distribution.

1.52: Factors Affecting the Permeability of Porous Materials

Compaction. Just as compaction reduces porosity, it also reduces the permeability of a porous material. Fibrous materials, such as paper, insulation materials, and wood suffer great reduction in permeability upon compaction. Unconsolidated materials, such as hard-grained powders and sands require relatively large compaction pressures to produce very significant reductions in permeability.

In the case of well-consolidated materials, such as sedimentary rocks, rather extreme compaction pressures are required to reduce the permeability significantly. However, for many materials, above a certain level of compaction pressure further increase in pressure does not significantly alter the permeability. This is shown in Figure 1-5 for some sedimentary rocks.

Clay Swelling. Many consolidated sandstones contain clays and silt to some degree. Since montmorillonite-type clays absorb fresh water to a considerable degree, with resultant swelling, the permeabilities of natura sandstones are generally greatly reduced when measured with fresh water

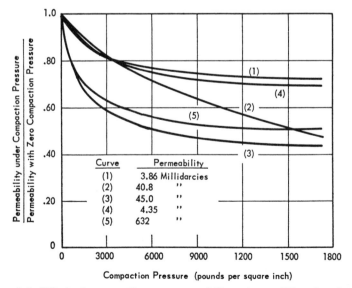

Figure 1-5. Effect of compaction on permeability of consolidated rocks. (*After Fatt and Davis, 1952.*)

The addition of salts, such as sodium chloride or potassium chloride, will, in most cases, eliminate this swelling of clays.

Leaching. Since calcium carbonate is relatively soluble in fresh water, the flushing of fresh water through a porous limestone tends to leach or dissolve the walls of the pores. This results in an increase in permeability. Such leaching effects can generally be eliminated by employing a saturated solution of calcium carbonate for permeability measurements.

Mechanical Alteration of Structure. Unconsolidated porous materials are subject to alterations in structure by mechanical forces acting on the grains or particles due to the flow of a viscous fluid through the structure. Such alteration may change the permeability.

1.60: Typical Values of Porosity, Permeability and Specific Surface

Table 1-1 gives typical values for the three basic characteristics of various types of porous materials. These examples serve to indicate the ranges in values which may be encountered for these basic physical characteristics.

1.70: Porosity and Permeability Distributions in Naturally Porous Materials

Since naturally porous materials possess a more or less random pore structure, it is not surprising that small samples of the same material do not have the same porosity or permeability. It is generally observed that the greater the volume, for the individual samples, the more likely will

TABLE 1-1. PHYSICAL CHARACTERISTICS OF TYPICAL POROUS MATERIALS

Substance	Porosity (fraction)	Specific surface (cm²/cm³)	Permeability (Darcys)	Reference
Silica powder	0.37–0.49	6.8×10^3–8.9×10^3	1.3×10^{-2}–5.1×10^{-2}	Carman 1938
Loose sand	0.37–0.50	1.5×10^2–2.2×10^2	20–180	Carman 1938
Soils	0.43–0.54	2×10^3–4×10^3*	29–140	Peerlkamp 1948
Sandstone	0.08–0.38	1.5×10^4–10×10^4*	5×10^{-4}–3.0	Muskat 1937
Limestone	0.04–0.10	0.15×10^4–1.3×10^4*	2×10^{-4}–4.5×10^{-2}	Locke & Bliss 1950
Brick	0.12–0.34	3×10^3–5×10^4*	4.8×10^{-3}–2.2×10^{-1}	Stull & Johnson 1940
Leather	0.56–0.59	1.2×10^4–1.6×10^4	9.5×10^{-2}–1.2×10^{-1}	Mitton 1945
Fiber glass	0.88–0.93	5.6×10^2–7.7×10^2	24–51	Wiggens *et al.* 1939

* Supplied by the author.

the same values be observed. This characteristic of porous materials can be understood by the following analysis.

Consider a great bulk of porous material and imagine it divided into very small rectangular parallelopipeds. These elements will possess a distribution of porosity due to the random porous structure. Let this distribution function be denoted by $F(\phi)$, so that $F(\phi)d\phi$ is the fraction of elements having porosity between ϕ and $\phi + d\phi$.

The mean porosity of these elements, and hence the actual porosity of the great bulk of material, is

$$\bar{\phi} = \int_0^1 \phi F(\phi) \, d\phi \tag{1-14}$$

Also the standard deviation, σ, of the distribution in ϕ is defined by

$$\sigma^2 = \int_0^1 (\phi - \bar{\phi})^2 F(\phi) \, d\phi \tag{1-15}$$

Now suppose that samples of the porous material are taken. Let the volume, V, of each sample be composed of n of the elemental parallelopipeds. The porosity of a sample is then

$$\phi_s = \frac{1}{n} \sum_{i=1}^n \phi_i \tag{1-16}$$

where the ϕ_i are the porosities of the elemental blocks.

According to the central limit theorem,[22] the distribution function for ϕ_s must approach, for large n, the Gaussian distribution. Thus, denoting the distribution function of ϕ_s by $G(\phi_s)$, we have by the central limit theorem

$$G(\phi_s) = \frac{1}{\sqrt{2\pi}\left(\dfrac{\sigma}{\sqrt{n}}\right)} \exp - \left\{ \frac{(\phi_s - \bar{\phi})^2}{2(\sigma/\sqrt{n})^2} \right\} \tag{1-17}$$

provided that $0 \ll \bar{\phi} \ll 1$. That is, the mean of the sample distribution is the same as the mean of the parent distribution of elemental blocks, and the standard deviation of the sample distribution is the standard deviation of the parent distribution divided by $\sqrt{n}$.

Since the number of elements, n, in a sample is proportional to the volume V, of the sample, it follows that the standard deviation observed for the samples is inversely proportional to the square root of the volume per sample. Thus, if we denote the volume of an elemental block by ϵ, we have

$$n = \frac{V}{\epsilon} \tag{1-18}$$

and therefore

$$\sigma_s = \sqrt{\frac{\epsilon}{V}} \, \sigma \tag{1-19}$$

where σ_s is the standard deviation for the sample distribution.

For very small elemental blocks ($\epsilon \to 0$), a unique distribution is obtained. As ϵ approaches zero, only two values of porosity are possible for the elemental blocks, namely 0 or 1. Furthermore, the fraction of blocks having zero porosity approaches $1 - \bar{\phi}$, and the fraction having porosity equal to 1 approaches $\bar{\phi}$. Thus, it can be expected that for some critically small value of ϵ, say ϵ_0,

$$\lim_{\epsilon \to \epsilon_0} \int_0^{\Delta \phi} F(\phi) \, d\phi = 1 - \bar{\phi} \tag{1-20}$$

and

$$\lim_{\epsilon \to \epsilon_0} \int_{1-\Delta \phi}^1 F(\phi) \, d\phi = \bar{\phi} \tag{1-21}$$

where $\Delta\phi$ is an infinitesimal increment in ϕ.

Then from the definition of σ, we have

$$\sigma^2 = \lim_{\epsilon \to \epsilon_0} \int_0^1 (\phi - \bar{\phi})^2 F(\phi) \, d\phi = \bar{\phi}(1 - \bar{\phi}) \tag{1-22}$$

and, therefore, the porosity distribution function for samples with volumes, $V \gg \epsilon_0$, is

$$G(\phi_s) = \sqrt{\frac{V}{2\pi\epsilon_0\bar{\phi}(1 - \bar{\phi})}} \, \exp \left\{ \frac{-V(\phi_s - \bar{\phi})^2}{2\epsilon_0\bar{\phi}(1 - \bar{\phi})} \right\} \tag{1-23}$$

The extent to which this equation is applicable to naturally porous materials has been only slightly investigated. Certainly, the only assumptions involved are that the porous structure is spatially random and that $V \gg \epsilon_0$.

The characteristic volume, ϵ_0, should be related to the size and uniformity of the pores and, consequently, should be related to permeability. This has been investigated for certain sandstones. Figure 1-6 shows two typical porosity distribution curves (porosity interval = 0.01) for the Woodbine sandstone of the Texas Gulf Coast. Figures 1-7 and 1-8 illustrate the observed dependence of the characteristic elemental volume, ϵ_0, and the average porosity, $\bar{\phi}$, on permeability.

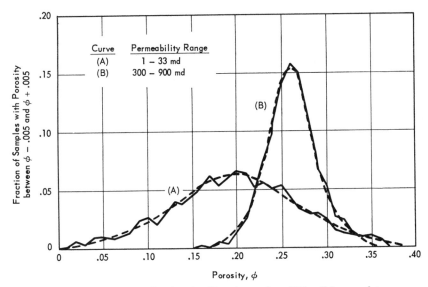

Figure 1–6. Porosity distribution for 20 cm³ samples of Woodbine sandstone, over 400 samples for each curve.

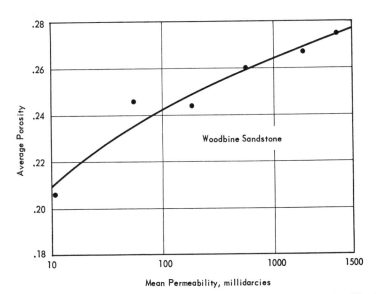

Figure 1–7. Correlation between mean porosity and permeability for Woodbine sandstone.

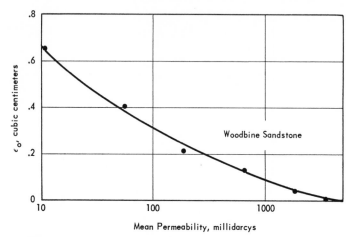

Figure 1–8. Distribution parameter ϵ_0 versus permeability for Woodbine sandstone.

A completely adequate theory describing the permeability distributions of naturally porous materials has not been devised. This is primarily because the measured permeability of a heterogeneous sample depends upon the direction of fluid flow employed in the measurement.

Several investigators have reported data of permeability distributions for sandstones. One such distribution is shown in Figure 1-9. An important

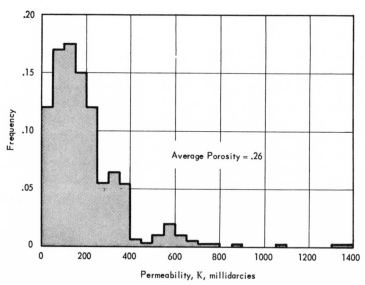

Figure 1–9. Permeability distribution for a natural sandstone. (*After Law, 1947.*)

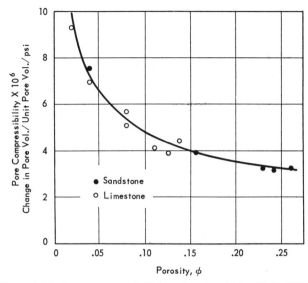

Figure 1–10. Pore compressibilities of rocks. (*After Hall, 1953.*)

characteristic of permeability distributions is their skewness. Thus, the most frequently occurring value of permeability is usually much closer to the harmonic mean of the distribution than to the arithmetic mean permeability. This is in direct contrast to the case of porosity, where the arithmetic mean porosity is the most frequently occurring value.

1.80: Mechanical Properties of Porous Materials

The mechanical properties of porous materials are not usually of importance in problems of fluid flow through such media. However, in the particular case of deeply buried sedimentary rocks, these properties may have some bearing on the flows of oil, water and gas.

In the petroleum industry several studies of rock compressibilities and rock strength have been made.

Compressibilities of Porous Rocks. Compressibility is defined by

$$c = -\frac{1}{V_B}\frac{\partial V_B}{\partial p} \tag{1-24}$$

where p denotes externally applied hydrostatic pressure.

For porous rocks, it is found that compressibility depends explicitly on porosity. This is shown in Figure 1-10 for a variety of rock types.*

* Pore compressibility and grain compressibility can be defined in a similar manner to bulk compressibility; as indicated in problem 3, these compressibilities are related. Actually, the separate effects of pore and confining pressures should be considered.

Figure 1–11. Failure of rock samples under compressive loading. *Left*, typical sample; *Center*, brittle failure; *Right*, malleable deformation with high confining pressure. (*After Robinson, 1959.*)

Compressive Strength. Studies conducted on porous limestones, sandstones and shales show that the state of stress of a porous rock greatly affects the compressive strength of the rock. In particular, differences in internal fluid pressure and external pressure on a sample determine the type of mechanical failure that will occur. As the excess of external over internal pressure increases, the mode of failure changes from brittle failure to malleable failure. The yield strength increases as this pressure difference is increased.[18]

Figure 1-11 illustrates the two types of failure: (a) is a typical rock sample, (b) is a sample which suffered brittle failure, and (c) is a sample which has been malleably deformed.

Some studies, particularly in the ceramic industry,[20] show that the manner in which consolidation of a porous material takes place determines to a major extent the mechanical strength of a consolidated porous material. In particular, when consolidation of a dry porous material is produced by recrystallization due to applied pressure, the compressive and tensile strength of the material increases with increasing consolidation pressure.

EXERCISES

1. From the definition of porosity, show that if a sample of porous material of volume V_T having porosity ϕ_T is cut into n pieces having volumes V_i, $i = 1, 2, \cdots, n$, then

$$\phi_T = \frac{1}{V_T} \sum_{i=1}^{n} V_i \phi_i$$

where the ϕ_i, $i = 1, 2, \cdots n$ are the porosities of the pieces. Also consider the case in which the V_i are all equal.

2. Show that for a cubic packing of spheres with radius R,

$$\Sigma = \frac{\pi}{2R}, \quad \phi = \frac{\pi}{6} \text{ and, minimum pore radius} = (\sqrt{2} - 1)R.$$

Compare the minimum pore radius with K/ϕ as given by the modified Kozeny equation. Use $\tau = 1.5$, and $c = 0.56$. (Note that for dimensions in centimeters the Kozeny equation gives K in cm^2.)

3. With the bulk, pore and solid compressibilities defined as

$$c_B = \frac{-1}{V_B}\frac{\partial V_B}{\partial p}, \quad c_p = \frac{-1}{V_p}\frac{\partial V_p}{\partial p} \text{ and } c_s = \frac{-1}{V_s}\frac{\partial V_s}{\partial p}$$

respectively, show that the definition of porosity implies

$$c_B = (1 - \phi)c_s + \phi c_p$$

where B, s and p refer to bulk, solid and pore quantities, respectively.

References

1. Athy, L. F., *Bull. Am. Assoc. Petroleum Geol.*, **14**, 1 (1930).
2. Brooks, C. S. and Purcell, W. R., *Trans. AIME*, **195**, 289 (1952).
3. Burdine, N. T., Gournay, L. S., and Reichertz, P. P., *Trans. AIME*, **189**, 195 (1950).
4. Carman, P. C., *J. Soc. Chem. Ind.*, **57**, 225 (1938).
5. Chalkley, J. W., Cornfield, J., and Park, H., *Science*, **110**, 295 (1949).
6. Darcy, H., "Les fontainer publiques de la ville de Dijon," Dalmont, Paris (1856).
7. Fatt, I., and Davis, D. H., *Trans. AIME*, **195**, 329 (1952).
8. Graton, L. C., and Fraser, H. J., *J. of Geol.*, **43**, 785 (1935).
9. Gregg, S. J., "The Surface Chemistry of Solids," Chapman & Hall, London, Reinhold Publishing, New York (1951).
10. Hall, H. N., *J. Petroleum Technol.* Tech. Note 149, Jan. (1953).
11. Kozeny, J., *S.-Ber. Wiener Akad. Abt. II a*, **136**, 271 (1927a).
12. Krumbien, W. C., and Monk, G. D., *Trans. AIME*, **151**, 153 (1943).
13. Law, J., *Trans. AIME*, **155**, 200 (1947).
14. Locke, L. C., and Bliss, J. E., *World Oil*, **131**, No. 4, 206 (1950).
15. Mitton, R. B., *J. Intern. Soc. Leather Trades Chem.*, **29**, 255 (1945).
16. Muskat, M., "Flow of Homogeneous Fluids through Porous Media," McGraw-Hill Book Co., New York (1937); J. W. Edwards, Inc., Ann Arbor (1946).
17. Peerlkamp, P. K., *Landbouwkund. Tijdschr.*, **60**, 321 (1948).
18. Robinson, L. H., *Trans. AIME*, **216**, 26 (1959).
19. Scheidegger, A. E., "The Physics of Flow through Porous Media," Macmillan Co., New York (1957).
20. Searle, H. B., and Grimshaw, R. W., "The Physics and Chemistry of Clays," Interscience Pub. Co., New York (1959).
21. Stull, R. T., and Johnson, P. V., *J. Res. Natl. Bur. Standards*, **25**, 711 (1940).
22. Uspensky, "Introduction to Mathematical Probability," McGraw-Hill Book Co., New York (1937).
23. Wiggens, E. J. et al., *Canadian J. Res.*, **317**, 318 (1939).

2. STATICS OF FLUIDS IN POROUS MEDIA

2.10: Fluid Saturations

The void space of a porous material may be partially filled with a liquid, the remaining void space being occupied by air or some other gas. Or, two immiscible liquids may jointly fill the void space. In either of these cases or the case of three immiscible fluids jointly filling the void space, the question as to how much of the void space is occupied by each fluid is very important.

The saturation of a porous medium with respect to a particular fluid is defined as the fraction of the void volume of the medium filled by the fluid in question. Thus, denoting the saturation with respect to fluid w by S_w, the definition of saturation is

$$S_w = \frac{\text{volume of fluid in the medium}}{\text{total volume of voids in the medium}} \qquad (2\text{-}1)$$

Thus for two fluids, w and nw say, jointly filling the void space, it follows that

$$S_w + S_{nw} = 1 \qquad (2\text{-}2)$$

with a similar relation holding for three immiscible fluids.

Observe that saturation is a bulk property which ignores the relative distributions of the fluids within the porous structure of the material. Also note that saturation is a dimensionless quantity.

2.11: Measurement of Fluid Saturations

Several methods have been rather widely used for measuring fluid saturations. These are described as follows.

Volumetric Balance Method. If a sample of porous material whose porosity is known is initially devoid of a fluid w and then a volume of the fluid, V_w, is introduced into the material, the saturation is calculated directly by conservation of volume. Thus

$$S_w = \frac{V_w}{\phi V_B} \qquad (2\text{-}3)$$

A similar procedure applies if the sample of porous material is initially saturated with the fluid and a volume V_w is withdrawn.

22

Direct Weighing. In the case of two immiscible fluids jointly saturating a porous medium the respective saturations can be determined by direct weighing. Thus, for example, if the weight of the porous material is determined in an evacuated (or gas-filled) state and again when partially saturated with a liquid of density ρ_l , the saturation with respect to the liquid is given by

$$S_l = \frac{W_2 - W_1}{\phi \rho_l V_B} \tag{2-4}$$

Here W_2 is the weight at liquid saturation S_l and W_1 is the weight when no liquid is present.

Electrical Resistivity Method. If a porous material is a poor conductor of electrical current then when the void space of the material is partially filled with a fluid which is a good conductor, such as a sodium-chloride solution, the saturation with respect to this fluid can be determined by electrical resistivity measurements.

This technique of saturation determination is based on Archie's Law which is discussed in section 2.60. It is particularly applicable in situations in which the fluid saturation is relatively uniform throughout the sample and weighing is impractical or impossible.

X-Ray Absorption Method. When x-rays traverse any material the intensity of the rays is attenuated in accordance with the exponential equation

$$I = I_0 e^{-\beta x} \tag{2-5}$$

Here I denotes intensity, I_0 the intensity of the beam for zero penetration distance, x is the thickness of material traversed and β is the absorption coefficient for x-rays in the material in question.

If, in a porous material jointly saturated by two immiscible fluids, one of the fluids contains a dissolved salt which is a good absorber of x-rays, then variations in the saturation of this fluid will be reflected in significant changes in the over-all x-ray absorption coefficient of the sample. Consequently, x-ray absorption can be used as a measure of saturation.[8, 13]

This technique is particularly applicable to studies of saturation distribution in two-phase flow problems, though it is not very precise.

2.20: Capillary Pressure

When two immiscible fluids are in contact a discontinuity in pressure exists between the two fluids which depends upon the curvature of the interface separating the fluids. This pressure difference, which we call the capillary pressure and denote by p_c , is given by Laplace's equation[7]

$$p_c = \gamma_{12} \left(\frac{1}{r} + \frac{1}{r'} \right)$$ (2-6)

Here r and r' are the principal radii of curvature of the interface and γ_{12} is the specific free energy of the interface.[9] Frequently, the specific free energy is interpreted as surface tension.

For two immiscible fluids in contact within a bounding solid surface, a capillary tube for example, the fluid-fluid interface intersects the solid surface at an angle termed the contact angle θ. This angle is determined by Young's equation[7]

$$\cos \theta = \frac{\gamma_{s1} - \gamma_{s2}}{\gamma_{12}}$$ (2-7)

γ_{s1} is the specific free energy of the interface between the solid and fluid number 1 and γ_{s2} is the corresponding quantity for the interface between the solid and fluid number 2.

Interfacial tension (or specific free energy of an interface) has the dimension of force per unit length. In most tables and handbook references, it is expressed in dynes per centimeter.

If $\gamma_{s1} > \gamma_{s2}$ then θ is an acute angle and fluid 1 is said to wet the solid. That is, fluid 1 has a greater tendency to spread over the solid surface than does fluid 2. For $\gamma_{s1} < \gamma_{s2}$ the converse is true.

The foregoing equations can be shown to be the consequence of the equilibrium requirement that the total free energy of a system in equilibrium be a minimum.[7]

If a porous material is completely saturated with fluid 2 and some of fluid 1 is introduced on its surface, $\gamma_{s1} > \gamma_{s2}$, then fluid 1 tends to flow spontaneously in along the walls of the pores, displacing fluid 2. The wetting fluid is said to displace the non-wetting fluid by imbibition. Equilibrium results when the wetting fluid has accumulated in those pores and interstices which permit the greatest curvature of the fluid-fluid interface, consistent with Young's equation. Thus, the wetting fluid tends to fill the smallest pores first.

The capillary equilibrium described can be most easily understood by considering a porous structure composed of cylindrical rods arranged in a parallel cubic packing as shown in Figure 2-1. Taking γ_{s2} to be 0 and $\gamma_{12} = \gamma_{s1}$, $\cos \theta$ is 1 and hence θ is 0. This corresponds to fluid 1 being water, fluid 2 being air and the rods being glass. For this case r' is infinite; that is, the fluid-fluid interface is a cylindrical sheet.

In cross section the rods and interfaces are as shown in Figure 2-2. The porosity of this simple structure is readily computed to be

$$\phi = 1 - \pi/4$$ (2-8)

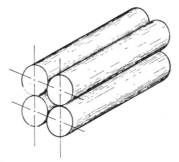

Figure 2-1. Cubic packing of circular rods.

Figure 2-2. Liquid-air interfaces in packing of glass rods.

and for interface radius, r, the saturation of fluid 1 is

$$S_1 = \frac{4}{3\pi} \left[\sqrt{\left(\frac{r}{R}\right)^2 + 2\frac{r}{R}} - \cos^{-1}\frac{R}{r+R} - \left(\frac{r}{R}\right)^2 \sin^{-1}\frac{R}{r+R} \right] \qquad (2\text{-}9)$$

where R is the rod radius.

The capillary pressure is simply

$$p_c = \frac{\gamma_{12}}{r} \qquad (2\text{-}10)$$

Thus, a relationship between capillary pressure and saturation has been obtained for this simple structure which holds until adjacent interfaces make contact. After the interfaces have made contact, this interface geometry becomes unstable. This relationship between capillary pressure and saturation is illustrated in Figure 2-3.

In naturally porous materials the porous structure is of an extremely complex random nature. Consequently, it is not possible to deduce the relationship between capillary pressure and saturation for such structures as was done for the simple model above. However, it is possible to measure capillary pressure at various saturations.

2.21: Capillary Pressure-Saturation Relationship

The displacement of one fluid by another in the pores of a porous medium is either aided or opposed by the surface forces of capillary pressure. As a

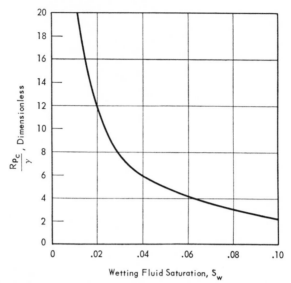

Figure 2–3. Computed capillary pressure-wetting fluid saturation relationship for water in cubic packing of glass rods.

consequence, in order to maintain a porous medium partially saturated with non-wetting fluid, while the medium is also exposed to wetting fluid, it is necessary to maintain the pressure of the non-wetting fluid at a value greater than that in the wetting fluid.

Denoting the pressure in the wetting fluid by p_w and that in the non-wetting fluid by p_{nw}, we have

$$p_{nw} - p_w = p_c(S_w) \tag{2-11}$$

That is, the pressure excess in the non-wetting fluid is the capillary pressure, and this quantity is a function of saturation. This is the defining equation for capillary pressure in a porous medium.

2.22: Measurement of Capillary Pressure

Gravity Method. The first method developed for determining the capillary pressure-saturation relationship of porous materials was developed for unconsolidated materials. This method is extensively employed in soil science.

Consider a vertical tube, filled with an unconsolidated porous material, having its lower end immersed in a basin of wetting fluid. The column is initially saturated with non-wetting fluid.

Taking the liquid level in the basin as zero elevation and measuring the vertical distance, z, from this point, the pressures in the two fluids at any height, z, are

$$p_w (z) = p_w (0) - \rho_w g z \qquad (2\text{-}12)$$

and

$$p_{nw} (z) = p_{nw} (0) - \rho_{nw} g z \qquad (2\text{-}13)$$

Here ρ_w and ρ_{nw} are the respective mass densities of the fluids and g is the acceleration of gravity. These equations must hold when hydrostatic equilibrium has been established. This may require a considerable period of time.

Subtracting the first equation from the second, and using the definition of capillary pressure (2-11), there results

$$p_c (z) = p_c (0) + (\rho_w - \rho_{nw}) \, g z \qquad (2\text{-}14)$$

But since at $z = 0$ the material is completely saturated with wetting fluid, $p_c (0)$ is zero and this becomes

$$p_c (z) = (\rho_w - \rho_{nw}) \, g z \qquad (2\text{-}15)$$

If, after equilibrium has been established, the column is quickly divided into small segments along its length, and the saturations of the pieces determined, the capillary pressure-saturation relationship is determined. (The modern technique is to employ a series of ring electrodes along the column and determine saturation by measuring electrical resistivities.)

Displacement Method. Capillary pressure can be measured by placing a sample of the porous material, saturated with wetting fluid, in a chamber filled with non-wetting fluid. The lower surface of the chamber, on which the sample rests, must be a semi-permeable plate, through which only wetting fluid can pass. Extending from this porous plate (which is also saturated with wetting fluid) is a graduated tube.

With the sample in this apparatus, the pressure of the non-wetting fluid in the chamber can be slowly elevated to some value and maintained. This causes some wetting fluid to be displaced from the sample. The amount displaced is read from the graduated tube. Since in this case p_w is maintained at atmospheric pressure, and p_{nw} and S_w are directly measured (S_w may be measured electrically), p_c can be computed. This can be repeated for a variety of values of p_{nw}, thus obtaining points defining a curve of p_c versus S_w.

Centrifuge Method. Although capillary pressures cannot be measured directly by the centrifuge method, this method can furnish data which can be converted, under certain circumstances, to a capillary-pressure curve.[19]

In the centrifugal capillary-pressure experiment a small uniform sample of the consolidated porous material, initially saturated with a wetting fluid, is rotated successively at a selected series of angular velocities and the quantity of the wetting fluid expelled at each velocity measured. The saturated sample is placed in a cup device containing a non-wetting fluid. During rotation both the fluid in the sample and the non-wetting fluid surrounding the sample are subjected to a centrifugal force which produces a pressure gradient directed outward from the axis of rotation. In the usual case the wetting fluid has a greater density than does the non-wetting fluid so that a higher pressure is developed in the fluid within the sample. This causes the wetting fluid to flow out from the sample at the outer radius, being simultaneously replaced by non-wetting fluid entering at the inner radius. At a constant rate of rotation an equilibrium saturation distribution is formed. This distribution is determined by the capillary pressure-saturation relationship characteristic of the material.

If it is assumed that in any cross section of the sample, normal to the radius of rotation, the saturation is uniform, then by considering the forces acting on the fluids within the sample and applying Newton's second law of motion the following approximate equation can be derived,[10, 19]

$$\bar{S}_w p_{c1} = \int_0^{p_{c1}} S_w(p_c) \, dp_c \tag{2-16}$$

Here $\bar{S}_w$ is the average saturation of wetting fluid in the sample and p_{c1} is the capillary pressure at the inner radius of rotation of the sample, which is given by

$$p_{c1} = \frac{\Delta\rho\omega^2}{2} (r_2^2 - r_1^2) \tag{2-17}$$

where

$\Delta\rho$ = density difference in fluids
ω = angular rate of rotation (radians/time)
r_2 = outer radius of rotation of sample
r_1 = inner radius of rotation of sample.

Differentiating equation 2-16, there results

$$\bar{S}_w + p_{c1} \frac{d\bar{S}_w}{dp_{c1}} = S_w(p_{c1}) \tag{2-18}$$

Since $\bar{S}_w$ is measured and p_{c1} can be computed from equation 2-17, one can plot $\bar{S}_w$ versus p_{c1} from the data and estimate from the resulting curve the value of $d\bar{S}_w/dp_{c1}$. Then these values can be inserted into equation 2-18 and $S_w(p_{c1})$ computed. This yields the data finally necessary to plot p_{c1} versus S_{w1}.

2.23: Capillary Hysteresis

In the foregoing discussions of methods of determining capillary-pressure curves it was stated in each case that the sample was to be initially saturated with either wetting or non-wetting fluid. Actually, some of the foregoing methods can be applied with either initial state. However, the capillary pressure-saturation curves obtained for the two initial states are not the same. This phenomenon is termed capillary hysterisis.

The two capillary pressure-saturation curves have been given specific names. The curve obtained beginning with the sample saturated with wetting fluid is called the drainage curve, and that beginning with the sample saturated with non-wetting fluid is called the imbibition curve. These two curves for a sandstone and a kerosene-water liquid system are shown in Figure 2-4. These are typical of all such curves.

This difference in the saturating and desaturating capillary-pressure curves is closely related to the fact that the advancing and receding contact angles of fluid interfaces on solids are different. Furthermore, it frequently happens, particularly with natural crude oil-brine systems, that the contact angle, or wettability may change with time. Thus, if a rock sample which has been thoroughly cleaned with volatile solvents is exposed

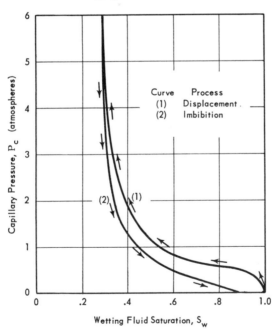

Figure 2–4. Typical capillary pressure-wetting fluid saturation relationship for a porous rock showing hysteresis.

to crude oil for a period of time, it will behave as though it were oil wet. But if, after cleaning, it is exposed to brine, it will appear water wet. At the present time one of the greatest unsolved problems in the petroleum industry is that of wettability of reservoir rocks.[3, 11]

Another mechanism which has been proposed to account for capillary hysterisis is the so-called "ink bottle" effect. This phenomenon can be easily observed in a capillary tube having variations in radius along its length.

Consider a capillary tube of axial symmetry having roughly sinusoidal variations in radius. When such a tube has its lower end immersed in water, the water will rise in the tube until the hydrostatic fluid head in the tube becomes equal to the capillary pressure.

If then the tube is lifted to a higher level in the water, some water will drain out, establishing a new equilibrium level in the tube.

When the meniscus is advancing and it approaches a constriction it "jumps" through the neck, whereas when receding it halts without passing through the neck. This phenomenon explains why a given capillary pressure corresponds to a higher saturation on the drainage curve than on the imbibition curve.

In most fluid-flow problems of practical interest, capillary hysterisis is not a serious problem because the flow regime usually dictates that one or the other capillary pressure-saturation curve will apply.

2.24: Irreducible Saturation (Connate Water)

All capillary pressure-saturation curves show a characteristically large slope for some low value of wetting fluid saturation. In most cases, the drainage capillary-pressure curve shows that extremely high (approaching infinite) pressures are required to produce an infinitesimal reduction in wetting fluid saturation when a particular limiting saturation is approached. This limiting saturation is called the irreducible saturation (or in the case of water, the connate water saturation).

Though the remaining fluid at the irreducible saturation can be removed (by heating for example), for all practical purposes it cannot be removed by injection of non-wetting fluid. Thus, in most physical problems, it will be assumed that p_c becomes infinite at a finite irreducible wetting fluid saturation.

For most porous materials a correlation exists between the irreducible saturation, S_c, and permeability. Such a correlation should exist since both quantities are related to "pore size." Figure 2-5 illustrates such a correlation for some samples of Woodbine sandstone from the Texas Gulf Coast.

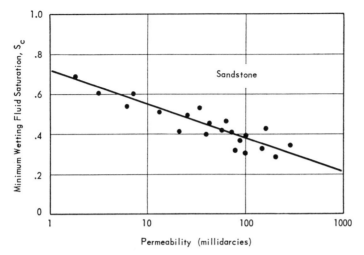

Figure 2–5. Relationship between minimum wetting fluid saturation (connate water saturation) and permeability for Woodbine sandstone.

2.25: The Leverett j-Function

The fact that the capillary pressure-saturation curves of nearly all naturally porous materials have many features in common has led to attempts to devise some general equation describing all such curves. Leverett[14] approached the problem from the standpoint of dimensional analysis.

Reasoning that capillary pressure should depend on the porosity, the interfacial tension and on some sort of mean pore radius, Leverett defined the dimensionless function of saturation which he called the j-function as

$$j(S_w) = \frac{p_c}{\gamma_{12}} \sqrt{\frac{K}{\phi}} \qquad (2\text{-}19)$$

In doing so he interpreted the ratio of permeability, K, to porosity, ϕ, as being proportional to the square of a mean pore radius.

This dimensionless capillary-pressure function serves quite well in many cases to remove discrepancies in the p_c versus S_w curves and reduce them to a common curve. This is shown for various unconsolidated sands in Figure 2-6. An extension of this correlation technique is discussed in section 9-3 in conjunction with model scaling techniques.

2.30: Pore-Size Distribution

Capillary pressure-saturation curves have been employed to infer what is termed the pore-size distribution of porous materials. If the porous

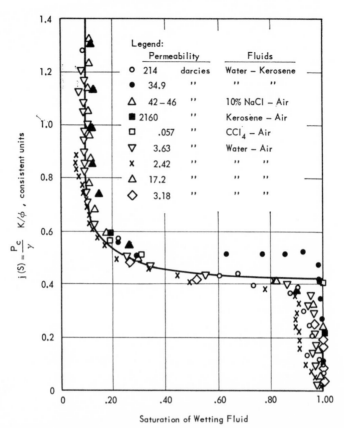

Figure 2-6. The Leverett j-function for unconsolidated sands. (*After Leverett, 1941.*)

medium is imagined as composed of a collection of "pores" having some distribution of radii, then equations can be derived which relate this distribution function to the capillary pressure-saturation function.[16] These equations are: the capillary pressure for a cylindrical tube of radius, r_i,

$$p_c(r_i) = \frac{2\gamma_{12} \cos \theta}{r_i} \tag{2-20}$$

and the relation between p_c and the distribution function, $D(r_i)$,

$$D(r_i) = \frac{p_c \phi}{r_i} \frac{dS_{nw}}{dp_c} \tag{2-21}$$

Here $D(r_i)dr_i$ is the fraction of the pore volume contributed by pores with radius between r_i and $r_i + dr_i$.

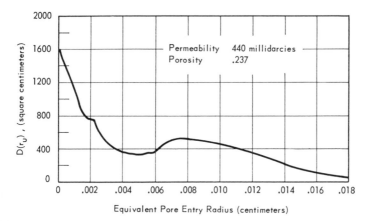

Figure 2–7. Typical pore size distribution for a sandstone as determined by the mercury-injection technique. (*After Burdine et al., 1950.*)

In these equations r_i is the "pore entry" radius. In particular, if mercury is injected into a porous medium the surface tension will give rise to an "entry pressure." Thus, if injection pressure is applied, the mercury enters the larger pores first and the pressure (capillary pressure) required to establish a certain saturation, S_{nw}, of this non-wetting fluid can be recorded. These data in conjunction with the above equations can be used to compute $D(r_i)$. Typical curves of this type are shown in Figure 2-7.

Pore size distribution curves such as these suffer a serious limitation which applies to all attempts to define pore size. In this procedure the pore entry is treated as cylindrical and certainly this is a gross simplification.[*] The complexity of naturally porous structures defies all attempts to define "pore size." Even so, simple definitions such as that employed above do have some utility and contribute to our understanding of the part played by pore size in determining the characteristics of flow through porous materials.

2.40: Vapor Pressure of Fluids in Porous Media

In studies of adsorption an important concept has been developed which has applications to the general theory of fluid behavior in porous materials. This concept is that of a force field[1] at the surface of a solid which acts on the molecules of liquid vapor. Thus, a wetting fluid is held within a porous material in two ways: one, by the force field which holds the liquid as a molecular film completely covering the surfaces of the porous structure,

[*] Recently Kruyer[12] has investigated this point. He has investigated an alternate procedure of treating a porous medium as a packing of spherical particles.

and two, by capillary forces which hold the liquid as bulk fluid with curved interfaces separating the liquid and vapor phases.

The properties of adsorbed fluid films have been the subject of considerable study.[1] In general, it appears that the properties of adsorbed liquid films are grossly different from those of the liquid in bulk. Such films exhibit properties more analogous to the solid phase, i.e. ice. Various theories have been developed to describe adsorption. These are reviewed by Scheidegger.[18]

For liquid held within a porous material in equilibrium with its vapor, an important factor appears. The partial pressure of vapor in equilibrium with bulk liquid depends upon the curvature of the vapor-liquid interface. Thus, since the curvature of the vapor-liquid interface in a porous medium depends upon saturation, it follows that the vapor pressure of a liquid in a porous material depends upon saturation. This dependence can be deduced as follows.

From thermodynamics, we know that the absolute free energy, F, of a fluid phase is a measure of the tendency for molecules to leave that phase. Thus, for a liquid and its vapor in equilibrium the free energy of the liquid, F_l, must be equal to that for the vapor, F_v. If a transition from one equilibrium state to another occurs, then we must have

$$dF_l = dF_v \tag{2-22}$$

Suppose then that the hydrostatic pressure in the liquid phase is changed by an amount dp. A corresponding change, dP, must occur in vapor pressure. If these changes occur under isothermal conditions, we have

$$dF_l = V_l dp \tag{2-23}$$

and

$$dF_v = V_v dP \tag{2-24}$$

where V_l and V_v are the specific molar volumes for liquid and vapor, respectively. Thus

$$V_l dp = V_v dP \tag{2-25}$$

If the vapor can be treated as an ideal gas, we can write

$$V_v = \frac{RT}{P} \tag{2-26}$$

Also, V_l can be expressed as

$$V_l = \frac{M}{\rho} \tag{2-27}$$

where M is the molecular weight of the liquid and ρ is the liquid density.

Substituting these expressions for V_l and V_v into equation 2-25, there results

$$\frac{dP}{P} = \frac{M}{\rho RT}\, dp \tag{2-28}$$

Now integrating with p going from atmospheric pressure, p_0, to a different pressure, p, and P going from a corresponding value P_0 to P, we obtain

$$P = P_0 \exp\left[\frac{M}{\rho RT}\,(p - p_0)\right] \tag{2-29}$$

When a wetting liquid is imbibed by a porous medium the difference in pressure between the vapor and the liquid is the capillary pressure

$$P - p = p_c \tag{2-30}$$

Thus

$$P = P_0 \exp -M/\rho RT(p_0 + p_c - P) \tag{2-31}$$

A further simplification can be made if the approximation

$$p_0 - P \approx 0 \tag{2-32}$$

is used. This is valid for

$$p_c \gg p_0 - P \tag{2-33}$$

Then

$$P \approx P_0 \exp - (M/\rho RT)p_c \tag{2-34}$$

which is the approximate relationship between vapor pressure within a porous medium and capillary pressure. Since capillary pressure is a function of the liquid (or vapor) saturation, it follows that the vapor pressure is a function of saturation.

Neither the exact nor approximate expressions can be expected to be strictly valid since the vapor does not behave exactly as an ideal gas. Furthermore, the above-described force field will play a dominant role at low fluid saturations. This is reflected in the dependence of p_c on saturation as well as in the dependence of P on saturation.

2.41: Measurement of Vapor Pressure-Saturation Curves

The vapor pressure-saturation relationship for a particular liquid in a porous medium can be determined by equilibrating a sample of the porous material in an atmosphere of fixed vapor pressure and measuring the liquid saturation of the sample. Such techniques have been intensively employed in soil science.[4]

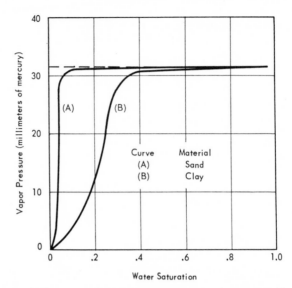

Figure 2–8. Typical curves of vapor pressure versus wetting fluid saturation.

Typical curves of P versus S_w obtained in this manner are shown in Figure 2-8.

2.50: Heat of Wetting and Heat of Swelling

When any dry substance is wetted several phenomena are observed. There is frequently a change in color, an increase in volume and an increase in temperature. Since heat is usually liberated at the same time that a change in volume takes place, considerable attention has been devoted to what has been called heat of swelling. This quantity is also closely related to the heat of wetting. The effects described here are most pronounced in soils, particularly clays. This is due to the presence of colloidal material.

The differential heat of swelling is defined as the quantity of heat liberated per unit change in volume. Thus

$$h_s = \left(\frac{\partial H}{\partial V_B}\right)_T \tag{2-35}$$

where h_s is the differential heat of swelling, H denotes quantity of heat liberated, V_B is bulk volume of porous material and the subscript T indicates that h_s is measured at constant temperature.

The relationship between the differential heat of swelling and the differential heat of wetting is determined by the rules of calculus as

$$\left(\frac{\partial H}{\partial V_B}\right)_T = \left(\frac{\partial H}{\partial V_L}\right)_T \left(\frac{\partial V_L}{\partial V_B}\right)_T \tag{2-36}$$

or

$$h_w = \left(\frac{\partial H}{\partial V_L}\right)_T = \left(\frac{\partial H}{\partial V_B}\right)_T \left(\frac{\partial V_B}{\partial V_L}\right)_T \tag{2-37}$$

Here V_L is the volume of liquid in the porous material and h_w is defined as the differential heat of wetting.

Defining the swelling coefficient as

$$\lambda = \left(\frac{\partial V_B}{\partial V_L}\right)_T \tag{2-38}$$

there results the relationship

$$h_w = \lambda h_s \tag{2-39}$$

The differential quantities defined above are difficult to measure directly but the corresponding integral quantities can be measured directly.

The integral heat of wetting is defined as the total heat liberated when a volume of liquid is added to dry material. Thus

$$H_w = \int_0^{V_L} h_w \, dV_L \tag{2-40}$$

The total heat of wetting $\bar{H}_w$ can be defined as the heat liberated when the liquid saturation is changed from zero to one hundred per cent.

The addition of the liquid to the porous material creates an area of new liquid-solid interface and destroys an equal area of air-solid interface. Since the respective free surface energies for these interfaces are γ_{SL} and γ_{SA}, the total energy gained in the process is

$$\bar{H}_w = (\gamma_{SL} - \gamma_{SA})\phi V_B \Sigma \tag{2-41}$$

Then, in view of Young's equation (2-7), this can be written as

$$\bar{H}_w = \gamma_{LA}\phi V_B \Sigma \cos\theta \tag{2-42}$$

We see that the total heat of wetting depends on the specific surface, the chemical nature of solid and liquid, and the porosity. It is easy to see from this relationship that clays, which have very great specific surfaces, should exhibit large total heats of wetting. For montmorillonite clay the total heat of wetting is about 15 to 20 calories per gram, whereas for most soils it is the order of 1 to 5 calories per gram.[4]

2.60: Electrical Properties of Fluid-Filled Porous Materials

Some of the electrical properties of fluid-filled porous materials are extremely important in studies of fluid flow. For example, electrical resistivity is used widely as a measure of brine saturation in rocks and unconsolidated sands. Those electrical properties of significance in studies of fluid flow through porous materials are described.

Formation Factor. The formation factor, F, is defined as the ratio of the resistivity* of the porous material saturated with an ionic solution to the bulk resistivity of the same ionic solution. (A sodium-chloride solution with over 10 g per liter of NaCl is usually used.) Thus

$$F = \frac{R_0}{R_w} \tag{2-43}$$

where R_0 is the resistivity of the saturated sample and R_w is the resistivity of the brine solution.

An empirical relationship between porosity and formation factor first pointed out by Archie[2] has been found to apply to most isotropic porous materials which do not contain electrically conducting solids. This relationship is

$$F = \phi^{-m} \tag{2-44}$$

where the exponent, m, has a value of approximately 2 for sands. For other materials m is generally less than 2.

The relationship found empirically by Archie is only statistically correct and has no theoretical basis. By considering a porous medium to be equivalent to a bundle of tortuous circular capillaries of equal size, a relationship is obtained which has been well substantiated by experiment. This is

$$F = \tau^n \phi^{-1} \tag{2-45}$$

where τ is the tortuosity defined as the ratio of the tortuous capillary length to the bulk length of the porous sample and n is a pure number.

Measurements of τ by an ion-migration technique have been carried out by Winsauer and co-workers.[20] Their values of τ, F and ϕ for a wide variety of porous rocks satisfy equation (2-45). Even so, this relationship is open to question because of the simple capillaric model employed.

However, τ is very difficult to measure and for practical purposes Archie's relationship is widely used.

Archie's Law. For porous materials partially saturated with a conduct-

---

* Resistivity is defined as the resistance of a cube having sides of unit length, measured with uni-directional current flow through one face and out the opposite face.

ing solution, which wets the solid, another empirical relationship, usually called "Archie's Law," exists. That part of the pore space not filled with the wetting-conducting solution is filled with non-wetting, non-conducting fluid. Archie's Law states that the resistivity of the partially saturated medium is related to S_w, the wetting-fluid saturation, by

$$R = R_0 S_w^{-n} \qquad (2\text{-}46)$$

Here R is the resistivity at S_w and n is a constant called the saturation exponent.

For clean sands n has a value of approximately two. For other materials n may be either greater than or less than two.

Typical plots of F versus ϕ, and R/R_0 versus S_w for sandstones are shown in Figures 2-9 and 2-10.

In general, the above relationships between porosity, saturation and electrical resistivity hold only approximately even for clean sands. For porous materials containing chemically active components, such as clays and shales, these laws must be significantly modified.[15]

Resistivity Measurements. The most widely used technique for measuring the resistivities of fluid-filled porous materials is the "four electrode

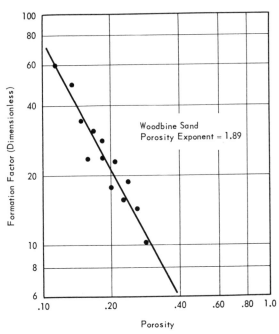

Figure 2–9. Correlation between formation factor and porosity for a sandstone.

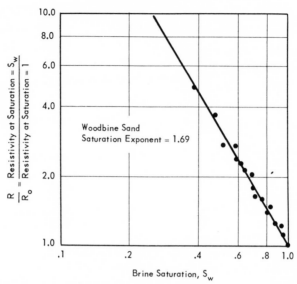

Figure 2–10. Typical correlation between resistivity ratio and wetting fluid saturation.

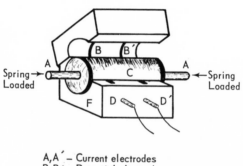

A,A′ – Current electrodes
B,B′ – Potential electrodes
C – Core sample
D,D′ – Contacts to potential electrodes
F – Hinged box core holder (insulator)

Figure 2–11. Apparatus for measuring electrical resistivity of cylindrical rock samples.

method." In this technique the sample, in the form of a short circular cylinder, is placed in a non-conducting jacket as indicated in Figure 2-11.

The current electrodes are in the form of plates placed on the ends of the sample, with contact being established by saturated sheets of blotting paper between plate and sample. The potential measuring electrodes are in the form of brass rings mounted on the inside walls of the jacket.

A square wave generator of about 60-cycles frequency is the best current supply. This avoids polarization, and nullifies ion separation.[17]

With this arrangement the resistivity, R, is given by

$$R = \frac{A}{L} \frac{V}{I} \qquad (2\text{-}47)$$

where A is the cross-sectional area of the sample, L is the distance between potential electrodes, V is the potential difference measured between these electrodes and I is the current.

2.70: Osmosis

A semi-permeable, or osmotic, membrane is a material through which fluids may pass but dissolved substances in these fluids will not pass. Generally if two solutions, one more dilute than the other, are separated by an osmotic membrane the solvent (usually water) will flow spontaneously through the membrane from the more dilute to the more concentrated solution.

In order to prevent this flow of solvent an applied pressure is required. This pressure is termed the osmotic pressure. For ideal membranes and ideal solutions, this pressure is determined by an equation identical to the ideal gas law. In particular, for an ideal solution having n moles of dissolved substance in a volume V of solution, separated from pure solvent by an ideal membrane the osmotic pressure is[9]

$$P = \frac{nRT}{V} \qquad (2\text{-}48)$$

where R is the gas constant and T is the absolute temperature. (Note that n is the number of moles in ions.)

Many porous materials behave as semi-permeable membranes, particularly clays and shales. However, such materials are not perfect osmotic membranes.

2.80: Electrochemical Potentials

Closely associated with the phenomenon of osmosis is that of electrochemical potentials.

A boundary potential results from the separation of electrical charge which occurs at the interface of two saline solutions of different concentrations when the speeds of migration of the positive and negative ions composing the salts are dissimilar.

This electrochemical potential is given approximately* for dilute solutions of binary electrolytes by[9]

* The resistivities replace activities.

$$E = \frac{RT}{ZF}\left(\frac{v-u}{v+u}\right)\ln\frac{R_1}{R_2} \tag{2-49}$$

where

R = gas constant
Z = valence of ions
F = the faraday
T = absolute temperature
v = anion mobility
u = cation mobility
R_1 = resistivity of solution 1
R_2 = resistivity of solution 2.

For sodium chloride at 25° C this becomes

$$E = 4.9934 \ln\frac{R_1}{R_2} \quad \text{(millivolts)} \tag{2-50}$$

These electrochemical potentials are modified when the solutions are separated by a chemically active porous material, such as clay or shale. The expression for solutions separated by such a membrane is approximately

$$E = C \ln\frac{R_1}{R_2} \tag{2.51}$$

where the value of C in general depends upon the nature of the membrane and the types of ions in the solution. For E in millivolts C has values ranging from about 5 to 25 millivolts.

2.90: Static Fluid Distributions in Porous Materials

The capillary and osmotic characteristics of fluids in porous media just described can be employed to deduce the static distribution of fluids in porous media arising in various physical problems. Examples of such problems are presented here.

Saturation Discontinuity at a Discontinuity in Medium. Consider two dissimilar porous materials in intimate contact, such as a layer of coarse sand overlaying a stratum of tight consolidated sandstone in the earth. Suppose the capillary pressure-saturation curves for the two sand bodies are as indicated in Figure 2-12. The upper curve, A, is that for the tight lower sand, while the lower curve, B, is that for the upper, unconsolidated sand.

Assuming that both sands contain some water and some air (and water

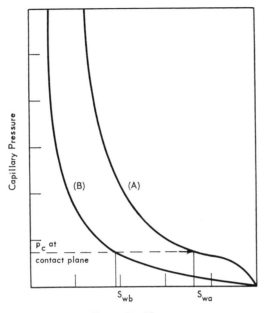

Wetting Fluid Saturation

Figure 2-12. Basis for discontinuity in saturation at a discontinuity in porous medium.

vapor), the saturations in the two sand bodies in the vicinity of their contact plane are to be determined.

From the definition of capillary pressure, we have

$$p_{cA} = p_{nwA} - p_{wA} \tag{2-52}$$

and

$$p_{cB} = p_{nwB} - p_{wB} \tag{2-53}$$

Since both the gas and liquid phases are continuous at the plane of contact, it follows that

$$p_{nwA} = p_{nwB} \tag{2-54}$$

and

$$p_{wA} = p_{wB} \tag{2-55}$$

at the plane of contact. Thus, we have

$$p_{cA} = p_{cB} \tag{2-56}$$

at the discontinuity in medium.

The horizontal line in Figure 2-12 connects two equal values of p_c on curves A and B. From this, it is seen that in order for this equality to obtain it is necessary for a discontinuity in saturation to exist at the plane of contact separating the two media, the magnitude of this discontinuity being $S_A - S_B$. Thus, for any water saturation in the loose sand less than one hundred per cent, a water saturation greater than this value will exist in the tight sand.

Shale Barriers, Salinity and Aquifer Pressures. An aquifer is a sand or rock stratum of the earth completely saturated with water (brine). Two distinct aquifers may be separated by a thin bed of shale or clay. If the salinities of the water in the two aquifers are unequal, then by osmosis water will percolate through the shale bed from the aquifer of lower salinity into the aquifer of higher salinity.

During this osmotic process, the salt ions in the dilute brine cannot move freely through the shale. Consequently, the salinity in this aquifer increases with time while that of the more concentrated aquifer decreases. As the salinities approach equality, the osmotic pressure decreases.

Equilibrium is attained when the difference in hydrostatic pressure between the aquifers becomes equal to the osmotic pressure across the shale bed.

Such osmotic phenomena can account for some of the observed water salinities and aquifer pressures observed in nature. This mechanism can also account for springs and other artesian flows observed at high elevations.

Distribution of Water in Surface Sands and Soil. The surface and near surface layers of soils, sands and clays of the earth's crust present a very complex problem with regard to the distribution of moisture. However, certain gross features of this distribution can be deduced from consideration of the properties of porous materials already discussed, provided that the investigation is restricted to an isothermal equilibrium state.

In order to simplify the problem as much as possible the discussion will be limited to fresh water, although the basic theory necessary for the inclusion of saline waters has been outlined.

Consider a multilayer system of sands, soils and clays arranged as horizontal strata of various thicknesses. For each layer a unique drainage capillary-pressure curve exists. It is assumed that equilibrium has been established by drainage.

Suppose that a water sand, i.e. a sand completely saturated with water, exists at a depth h. Since equilibrium exists in the earth layer overlaying this sand, we have

$$p_{c1} = \rho_w g(h - z) \tag{2-57}$$

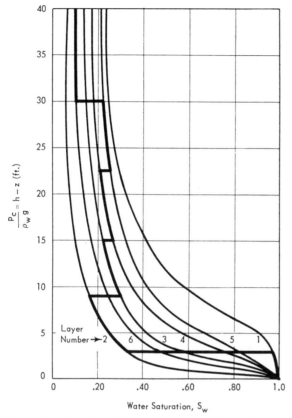

Figure 2-13. Determination of moisture distribution in surface sands and soils.

where ρ_w is the density of water, g is the acceleration of gravity and $(h - z)$ is the height above the water sand. Here p_{c1} denotes the capillary pressure in the first earth layer above the water sand.

The earth layers above the water sand are numbered 1, 2, 3, etc. consecutively upward and have corresponding thicknesses T_1, T_2, etc.

Now plot all the capillary-pressure curves on one graph as

$$\frac{p_c}{\rho_w g} = h - z \quad \text{versus} \quad S_w \tag{2-58}$$

Such a plot is shown in Figure 2-13. Then one can trace directly the composite representation of $h - z$ versus S_w by noting, for example, that curve 1 is followed until $h - z = T_1$. At this point a discontinuity occurs and curve 2 is followed until $h - z = T_1 + T_2$, etc. Proceeding in this manner the heavily lined composite curve is obtained. This composite curve represents the water saturation versus height above water-table curve, that is, the moisture distribution.

At the surface, here taken as 40 feet above the water table, vapor equilibrium between soil moisture and the atmosphere must obtain. Thus, according to equation (2-34), the temperature T must be adjusted.

EXCERCISES

1. Derive equation (2-21).
2. A thin stratum of porous sandstone is inclined at an angle θ to the horizontal. A thin horizontal layer of shale divides the stratum into two portions. Suppose the shale layer is a perfect osmotic membrane of 30 cm thickness and the pore space of the sandstone filled with saline (NaCl) water of ionic concentration 0.3 mol/liter above the shale and 0.1 mol/liter below the shale. Calculate the static pressure distribution in the sand assuming no flow.
3. Of the various types of surface soils, which hold moisture most effectively? Why? Considering gravity drainage to lower layers and evaporation to the atmosphere which mechanism would usually dominate in drying of surface soil?

References

1. Adams, N. K., "The Physics and Chemistry of Surfaces," Oxford Univ. Press, London (1941).
2. Archie, G. E., *Trans. AIME*, **146**, 54 (1942).
3. Bobek, J. E., Mattax, C. C., and Denekas, M. O., *Trans. AIME*, **213**, 155 (1958).
4. Barer, L. D., "Soil Physics," Chapman and Hall, London; John Wiley & Sons, New York (1940).
5. Bond, R. L., Griffith, M., and Maggs, F. A. P., *Discussions Faraday Soc.*, No. 3., 29 (1948).
6. Burdine, N. T., Gournay, L. S., and Reichertz, P. P., *Trans. AIME*, **189**, 195 (1950).
7. Collins, R. E., and Cooke, C. E., *Trans. Faraday Soc.*, **55**, 1602 (1959).
8. Geffen, T. M., and Gladfelter, R. E., *Trans. AIME*, **195**, 322 (1952).
9. Glasstone, S., "Textbook of Physical Chemistry," D. Van Nostrand Co., New York (1946).
10. Hessler, G. L., and Brunner, E., *Trans. AIME*, **160**, 114 (1945).
11. Holbrook, O. C., and Bernard, G. G., *Trans. AIME*, **213**, 261 (1958).
12. Kruyer, S., *Trans. Faraday Soc.*, **54**, 1758 (1958).
13. Laird, A. D. K., and Putnam, J. A., *Trans. AIME*, **192**, 275 (1951).
14. Leverett, M. C., *Trans. AIME*, **142**, 152 (1941).
15. Patnode, H. W., and Wyllie, M. R. J., *Trans. AIME*, **189**, 47 (1950).
16. Ritter, L. C., and Drake, R. L., *Ind. Eng. Chem.*, An. Ed. **17**, 782 (1945).
17. Rust, C. B., *Trans. AIME*, **195**, 217 (1952).
18. Scheidegger, A. E., "The Physics of Flow through Porous Media," Macmillan Co., New York (1957).
19. Slobod, E. L., Chambers, A., and Prekn, W. L., *Trans. AIME*, **192**, 127 (1951).
20. Winsauer, W. O., Shearin, H. M., Jr., Mason, P. H., and Williams, M., *Bull. Am. Assoc. Petroleum Geol.*, **36**, 253 (1952).

3. PHYSICAL AND MATHEMATICAL THEORY OF FLOW

3.10: Mechanisms and Types of Flow

Several mechanisms of fluid flow through porous materials are known to exist. The primary mechanism is, of course, of a purely "mechanical" nature, namely, flow as a result of an applied force in the form of a pressure differential. However, flow may also occur under certain circumstances as a result of applied electrical or thermal gradients.

Since mechanical flow, or flow induced by externally applied pressures, may be of several different types depending upon the range of pressure difference, average pressure, pore sizes, etc., it is necessary to discuss *types of flow* as well as *mechanisms of flow*.

3.11: Laminar Viscous Flow Through Porous Media

Darcy's Law. Laminar flow of a fluid is characterized by a fixed set of streamlines. A fluid element which at one point is traversing the same path as another element must follow the path of this element throughout its course. This is in contrast to turbulent flow in which only partial correlation between particle paths exists.

The viscosity of a fluid is a measure of internal friction associated with laminar flow. Shear forces exist apparently between lamellae of fluid having different velocities. An *ideal* viscous fluid flowing over a solid surface adhers to the surface. At the surface of the solid the fluid velocity is zero. As a result of this fluid sticking and the viscosity of the fluid, a *drag* force is exerted on the solid by the fluid. The fluid tends to drag the solid along with it.

Conversely, if the solid is held fixed, a force opposing the fluid motion is exerted on the fluid by the solid. This viscous resistance is a force equal and opposite to the drag force on the solid.

For a flat plate, the shear force per unit area between the solid surface and a fluid tangent to it is given by Newton's equation

$$F = \mu \left(\frac{dv}{dz}\right)_{\text{solid}} \tag{3-1}$$

Here μ is the fluid viscosity, v is fluid velocity which is a function of position above the plate, and z is distance from the surface into the fluid. The derivative is evaluated at the surface of the solid.

Since fluids have mass it follows from Newton's second law of motion that forces must be exerted on a fluid to change either the direction or magnitude of the fluid velocity. When a fluid flows through a porous medium the velocity of a fluid element changes rapidly from point to point along its tortuous flow path. The forces which produce these changes in velocity vary rapidly from point to point.

However, in a naturally porous material the porous structure and hence the multitude of flow paths have a random character. It is reasonable to suppose that the random variations in flow path for any particular fluid element are uniformly distributed. Also the variations in magnitude of velocity can be expected to be distributed uniformly with mean zero. Thus, for steady laminar flow the lateral forces associated with the microscopic random variations in velocity can be expected to average to zero over any macroscopic volume. However, the inertial forces in the direction of flow will not average to zero and hence will only be negligible for low flow rates. (See section 3.12.)

The only non-zero macroscopic force exerted on the fluid by the solid is that associated with the viscous resistance to flow. For steady laminar flow, this force must be in equilibrium with the external and body forces on the fluid. To formulate this equilibrium condition consider the flow apparatus shown in Figure 3-1.

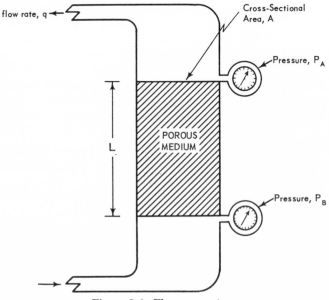

Figure 3-1. Flow apparatus.

A cylindrical sample of porous material having parallel ends, cross-sectional area A, and length L is mounted in a tube. The walls of the tube are tightly bonded to the sample. It is supposed that the system is filled with fluid.

For flow upward through the sample a net viscous resistance directed downward opposes the flow. In principle, this force could be computed by integrating equation (3-1) over the internal pore surface. Since the geometry of the pore surface is beyond mathematical description, this cannot be done. Even so, certain facts about this force can be stated mathematically.

Since for laminar flow the *relative* distribution of velocity within the pores is independent of the magnitude of the velocity, it follows that v and hence dv/dz must be everywhere proportional to q/A where q is volumetric flow rate. Furthermore, since the total surface involved must be proportional to the bulk volume, AL, of porous material, it follows that the viscous drag on the fluid can be written as

$$F_\mu = B\mu qL \tag{3-2}$$

B is a constant with dimensions of reciprocal length squared which is characteristic of the pore geometry.* The force F is directed downward for upward flow.

The external forces acting on the fluid contained within the porous sample can be expressed in terms of the pressures p_a and p_b at the ends of the sample. Since the pore areas on which these pressures act are given by ϕA, where ϕ is the porosity of the sample, the net upward force on the fluid due to these pressures is

$$F_p = (p_b - p_a)\phi A \tag{3-3}$$

The body force on the fluid is simply the weight of the fluid in the sample. This corresponds to a downward force

$$F_g = \rho(\phi AL)g \tag{3-4}$$

where ρ is the mass density of the fluid and g is the acceleration of gravity.

For steady flow, the forces F_μ, F_p and F_g must be in equilibrium. Thus

$$B\mu qL + \rho(\phi AL)g = (p_b - p_a)\phi A \tag{3-5}$$

or

$$q = -\frac{KA}{\mu L}[(p_a - p_b) + \rho gL] \tag{3-6}$$

where

* Note that B is proportional to the specific surface.

$$K = \frac{\phi}{B} \tag{3-7}$$

is a constant characteristic of the porous medium.

The constant K is the permeability of the porous medium discussed in section 1.50 and equation (3-6), which constitutes the operational definition of K, is Darcy's law. Actually, this is an empirical law[4] and the "derivation" given here is intended only as a heuristic guide to the understanding of its physical content. It should be noted that, for horizontal linear flow, gravitational effects do not come into play and the term $\rho g L$ must be omitted in equation (3-6). This corresponds to the case employed in the discussion of permeability in section 1.50.

The derivation of Darcy's law could lead one to think that it applied only for steady flow. Actually, the viscous forces involved in laminar flow through porous media are so much greater than any inertial forces associated with even relatively rapid variations in flow rate that inertial factors can in nearly all practical cases be neglected. Thus, for practical purposes Darcy's law is also valid for variable rate, q.

Darcy's Law for Gases. Darcy's law of laminar flow in the form of equation (3-6) is also valid for gases provided the flow rate, q, is taken as the volumetric flow rate as measured at the mean pressure $(p_a + p_b)/2$, and provided this mean pressure is sufficiently large. Writing $\bar{q}$ for the flow rate of the gas measured at mean pressure $(p_a + p_b)/2$ and q_a for the flow rate measured at pressure p_a, it follows from the ideal gas law that for isothermal conditions:

$$\bar{q} \frac{p_a + p_b}{2} = q_a p_a \tag{3-8}$$

Thus Darcy's law for ideal gases can be written as

$$q_a p_a = -\frac{KA}{\mu L} \left[\frac{p_a^2 - p_b^2}{2} + \left(\frac{p_a + p_b}{2} \right)^2 \frac{M}{RT} g L \right] \tag{3-9}$$

Here the density, ρ, is also obtained from the ideal gas law as

$$\rho = \frac{M}{RT} p = \frac{M}{RT} \frac{p_a + p_b}{2} \tag{3-10}$$

Of course, gravitational effects are essentially negligible for gases so g can be put equal to zero in most cases.

Actually, in the case of gases the fluid does not stick to the walls of the pores as required in Darcy's law and a phenomenon termed *slip* occurs. This slipping of the fluid along the pore walls gives rise to an apparent dependence of permeability on pressure. This dependence was pointed out by Klinkenberg[15] and is usually called the Klinkenberg effect. The relation between permeability and pressure proposed by Klinkenberg is

$$K = K_\infty \left(1 + \frac{b}{p}\right) \tag{3-11}$$

where K_∞ is the permeability as observed for incompressible fluids (liquids), p is the mean flowing pressure and b is a constant characteristic of both the gas and the porous medium.

From this equation, it is evident that for sufficiently high flowing pressures slip can be neglected and equation (3-9) applies with K being replaced by K_∞. When such is not the case, the flow is called *slip flow* and Darcy's law as modified by Klinkenberg must be employed. Thus, for slip flow

$$q_a p_a = -\frac{K_\infty A}{\mu} \left(1 + \frac{2b}{p_a + p_b}\right) \left(\frac{p_a^2 - p_b^2}{2L}\right) \tag{3-12}$$

where gravitational effects are considered negligible.

At very low pressures the flow process reduces to essentially a diffusional process. At very low molecular density gas flow becomes molecular streaming. Flow phenomena of this type are discussed by Barrer.[1]

The Measurement of Permeability. Permeability, K, is measured by establishing linear flow through the sample and applying the appropriate form of Darcy's law. Most frequently, gas flow is employed. Measurements at several mean pressures are necessary. Then these values can be plotted versus $2/(p_a + p_b)$. This plot yields a straight line with slope bK_∞ and intercept K_∞. When liquids are used, a single value K, equivalent to K_∞, is obtained.

The main precaution to be observed in such measurements is in preventing bypass flow around the sample.

Techniques for special types of permeability measurements are discussed in sections 4.20 and 4.40.

3.12: Turbulent Flow through Porous Media

The laminar flow regime breaks down for sufficiently high flow rates, q. For high flow rates, Darcy's law is not valid. The range of flow rate for which laminar flow exists has been studied by numerous investigators.[5, 7] Generally, this range is defined in terms of the Reynolds number. For example, in sands and sandstones, the transition from laminar to turbulent flow occurs rather gradually in the range of Reynolds number from one to ten.* The Reynolds number is defined as

$$R = \frac{q\rho\delta}{\mu A\phi} \tag{3-13}$$

where q is volumetric flow rate, ρ is fluid density, μ is viscosity, ϕ is porosity, A the cross-sectional area of the porous sample and δ is the average sand-grain diameter.

Usually, as in flow through pipes, for example, a dimension of the flow channel is employed in defining the Reynolds number. However, for porous media it is difficult to define, in terms of measurable quantities, a channel diameter. The grain diameter is usually employed as a measure of pore diameter. An alternate possibility is to employ $(K/\phi)^{1/2}$, with K expressed in cm^2, as a measure of pore diameter.

By analogy to flow through pipes, the flow of fluids through porous media can be studied in both laminar and turbulent states by considering the correlation of Reynolds number, R, and "friction factor," λ. The friction factor is defined as

$$\lambda = \delta \frac{\Delta p}{L\rho} \left(\frac{\phi A}{q} \right)^2 \tag{3-14}$$

Here Δp is the pressure differential across the length of the sample, L.

For the Reynolds number less than unity, the correlation between R and λ is found to be

$$\lambda = cR^{-1}$$

where c is a constant. This yields

$$q = \frac{\phi\delta^2 A}{c\mu} \frac{\Delta p}{L} \tag{3-15}$$

which corroborates Darcy's law. In fact, if $(K/\phi)^{1/2}$ is used in lieu of δ in the definitions of R and λ the constant c turns out to be unity.

Due to the fact that in porous media a distribution of pore sizes exists, the transition from laminar to turbulent flow is not abrupt at a critical Reynolds number as is the case for flow through pipes. Instead the transition is rather gradual.

An alternate procedure to empirical correlations between λ and R is to base the law for transition and turbulent flow on analogy to flow through tubes. Thus, Forchheimer[10] proposed an equation of the form

$$\frac{\Delta p}{L} = a\left(\frac{q}{\phi A} \right) + b\left(\frac{q}{\phi A} \right)^2 \tag{3-16}$$

---

* Strictly speaking, this transition region represents the onset of inertial effects. True turbulence occurs only at the higher Reynolds numbers.

where a and b are constants depending on the properties of both fluid and porous media. An equation of this form describes experimental results rather well for sufficiently high flow rates.*

Fortunately, the flow regime in most cases of practical interest is of the slow laminar type and Darcy's law applies. The mathematical theory of flow through porous media is generally formulated with Darcy's law being taken as the fundamental law of flow.

3.13: Simultaneous Laminar Flow of Immiscible Fluids Through Porous Media

Relative Permeability. Two immiscible fluids, such as an oil and water, may flow simultaneously through a porous medium. Experimentally, this can be accomplished by mounting a cylindrical sample of porous material within a tube so that the wall of the tube is tightly bonded to the sample. On each end a mixing and dispensing plate is mounted. The two fluids are introduced simultaneously through one plate and flow out from the other plate. The plates assure the uniform entry of the fluids over the entire cross section of the sample.[17]

Flow experiments conducted with such apparatus show that the permeability concept, and Darcy's law, can be extended to such multiphase flow. That is, a permeability can be defined for each fluid and Darcy's law then describes the relationship between the flow rate of each fluid and the pressure differential. The important fact in this is that these permeabilities are independent of flow rate and fluid properties. In fact, these permeabilities depend only on the fluid saturation within the porous sample.[11] Thus, with $K_1(S_w)$ and $K_2(S_w)$ denoting the permeabilities for the two fluids as functions of wetting-fluid saturation, equation (3-6) with the viscosity, μ, and density ρ of the fluid in question, can be written for the flow rates, q_1 and q_2, of the two fluids.

The permeabilities, K_1 and K_2, can be expressed as fractions of the single-phase permeability, K, of the porous medium. If, instead of the numerical subscripts, the subscripts w and nw for wetting and non-wetting fluid respectively are employed, then relative permeabilities are defined as

$$k_{nw} = K_{nw}/K \tag{3-17}$$

and

$$k_w = K_w/K \tag{3-18}$$

These relative permeabilities are each less than unity. However, their sum is not unity. Typical relative permeability curves are shown in Figure

---

* The analysis of flow through a non-uniform capillary tube given in Chapter 9 indicates that an equation of this form should describe flow including inertial effects.

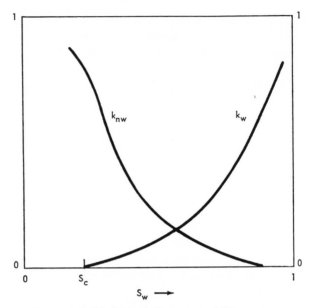

Figure 3-2. Typical relative permeability curves.

3-2. Here it can be seen that $k_w + k_{nw}$ is less than unity for all saturations, except possibly zero and one.

While relative permeabilities are essentially independent of fluid properties and flow rate in the laminar region, it is always necessary to distinguish between the wetting and non-wetting fluids.

Chatenever and Calhoun[3] have made visual studies of two-phase flow through packings of small glass and/or Lucite beads. The beads were packed between glass plates and the flow was observed with a microscope. The fluids used were oil and water, with one fluid being colored.

Such studies show that, when immiscible fluids flow simultaneously through a porous medium, each fluid establishes its own tortuous channels of flow through the medium. These channels are very stable and no turbulence or eddies are observed. A unique set of channels appears to occur for every range of saturation.

As the saturation of the non-wetting fluid is reduced, the channels for this fluid tend to break down until isolated islands of non-wetting fluid remain. These islands are stationary for all reasonable pressure gradients (laminar flow). The saturation of non-wetting fluid corresponding to this immobile state is, in the case of oil-water systems, called the residual oil saturation. Figure 3-3 shows the form of a residual oil globule in a sand. Generally, pressure gradients of the order of several atmospheres per

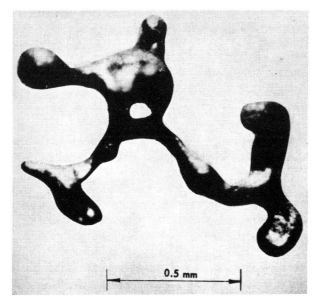

Figure 3–3. Residual oil globule. (*After Jordan et al., 1957.*)

centimeter are required to reduce the residual oil saturation. Such gradients correspond to turbulent flow.

Similarly, as the saturation of wetting fluid is reduced, the channels of flow for this fluid tend to break down and become discontinuous. The wetting fluid ceases to flow. The wetting-fluid saturation corresponding to this case is for water-oil systems called the connate-water saturation. (Cf. section 2.7.)

In the case of a gas-liquid system the slippage phenomenon noted for gas flow alone also occurs. Estes and Fulton[6] have shown that the permeability to gas in such flow regimes exhibits the Klinkenberg effect. As the mean flowing pressure is increased, the relative permeability curves approach those corresponding to an oil-water system.

The understanding of relative permeability and also capillary pressure saturation relationships has been greatly advanced by the work of Fatt.[8] Fatt has shown that these relationships, as well as other characteristics of porous media, can be understood in terms of a random network of simple capillary tubes.

The Measurement of Relative Permeabilities. Relative permeabilities can be measured by employing simultaneous linear flow of immiscible fluids. However, a detailed discussion of such measurements is best deferred until the mathematical theory of such multiphase flow has been developed.

Relative permeability measurements are discussed in section 6-10 in conjunction with the boundary effect and in section 6-21 in conjunction with the Welge integration of the Buckley-Leverett equation.

3.14: Streaming Potential and Electro-Osmosis

When two chambers containing water are separated by a plate of porous material and a difference of electrical potential between the chambers is maintained with a battery or power supply by two metal electrodes placed on each side of the porous plate, a flow of water occurs from one side to the other. This phenomenon is called *electro-osmosis*. Liquids other than water also exhibit this behavior.

Conversely, when water or other liquid is forced through a porous plate by a pressure differential an electrical potential difference across the plate is observed. This potential is called the *streaming potential*.

Both of these phenomena are associated with the existence of charged layers at the solid-liquid interfaces within the porous plate. Such charged layers are generally present at an interface separating two phases. Actually, these charged layers are diffuse in nature but by treating them as discrete layers a simple mathematical treatment can be applied to elucidate the basic features of the electrokinetic phenomenon described above.

The diffuse charged layers at a solid-liquid interface behave as two parallel surfaces of opposite electrical charge separated by a distance of molecular dimensions. A layer of one charge is on the solid surface and a layer of opposite charge in the fluid adjacent to the solid surface. The potential difference between these layers of charge is called the *zeta potential*.

If the charged double layers are treated as a condenser with parallel plates, a distance d apart the zeta potential, ζ, can be related to the charge per unit area, e, on a plate. If the dielectric constant of the medium between the plates is D, then from electrostatics*

$$\zeta = \frac{ed}{D\epsilon_0} \text{ (volts)} \tag{3-19}$$

When an electrical potential is applied across the porous medium a potential gradient, E, exists within the fluid-filled pore space. Denoting the component of E parallel to the pore surface by E_a, the force tending to displace the charged layer of fluid parallel to itself is eE_a per unit surface area. As the electrical force displaces the charges, the fluid is dragged along with the charges.

The total displacement force on the fluid within the porous plate is given

---

* M.K.S. units with: $\epsilon_0 = 8.85 \times 10^{-12}$ (coulombs)^2/newton-(meter)^2.

by the integral of eE_a over the pore surface. As in the case of the viscous drag, this cannot be evaluated since the pore geometry cannot be mathematically described. However, e should every. here be the same and E_a should everywhere be proportional to V/L where V is the potential difference across the porous plate and L is the plate thickness. Furthermore, the total force should be proportional to the total pore surface area, which, in turn, is proportional to the bulk volume, AL, of the porous plate. Thus. the displacement force can be written as

$$F_E = CeAV \qquad (3\text{-}20)$$

where C is a constant, having dimensions of reciprocal length, which is characteristic of the porous material.

For steady flow the displacement force, F_E, and the viscous force, F_μ, opposing flow must be in equilibrium. Thus, from equation (3-2)

$$CeAV = B\mu qL \qquad (3\text{-}21)$$

which neglects gravity forces. Then,

$$q = \frac{C}{B}\frac{e}{\mu} A \frac{V}{L} \qquad (3\text{-}22)$$

But since $B = \phi/K$, this can be written as

$$q = \frac{KeC}{\phi\mu} A \frac{V}{L} \qquad (3\text{-}23)$$

or, in view of equation (3-19)

$$q = C\epsilon_0 \frac{K\zeta D}{d\phi\mu} A \frac{V}{L} \qquad (3\text{-}24)$$

The direction of flow relative to the direction of the applied potential difference, V, will depend upon the sign of the charged layer within the fluid. In most cases, the solid surface is negatively charged. This, of course, depends on the nature of the fluid and the solid surface.

Equation (3-24) gives the magnitude of the electro-osmotic flow rate. Actually, this equation also gives the magnitude of the streaming potential arising from fluid flow. Thus, if, due to an applied pressure differential, a flow rate q is produced, then a potential difference, V, as given by equation (3-24) would be produced across the porous plate. This is so because as the fluid is mechanically displaced, it carries the charged fluid layer with it and this flow of charge constitutes an electrical current.

While the factors c and e are only qualitatively defined in the above discussion, the resulting equation shows how electro-osmosis and streaming potential are related to certain properties of fluid and solid.

Generally, electrokinetic phenomena are of little significance in most problems of fluid flow through porous media. The magnitudes of these electrokinetic effects are discussed by Glasstone.[12]

3.15: Thermo-Osmosis

Another mechanism of fluid transport through porous media is that produced by a temperature gradient. This phenomenon occurs only when the porous medium is just partially saturated with a liquid, the remaining portion of the pore space is filled with a gas and liquid vapor.

If, under such conditions, a temperature gradient is applied across the porous medium, it is observed that a movement of liquid from the warmer to cooler regions occurs. Several mechanisms for this phenomenon have been proposed: vaporization and recondensation of liquid and hence motion of the vapor phase; viscous flow of liquid due to the gradient of vapor pressure; and viscous flow of liquid due to a gradient of capillary pressure resulting from the temperature dependence of surface tension.

Studies of this phenomenon have been reported in the literature,[16, 19] but apparently a complete quantitative theory is still lacking.

3.20: Limitations of Classical Hydrodynamics

Classical hydrodynamics describes the flow of ideal viscous fluids within prescribed boundaries. In order to apply the mathematical equations of this theory to a particular fluid-flow problem, it is necessary that the boundaries of the system in question be mathematically described. In the case of flow through natural porous media, the complexity of the porous structure prohibits such description. Thus, if a mathematical theory of flow through porous media is possible at all, it must take the form of a statistical theory, or a theory based upon laws describing only the macroscopic features of the flow. The latter course has proved to be not only possible but also very successful.

3.30: Differential Form of Darcy's Law

Homogeneous Incompressible Fluids. In section 3.11 Darcy's law of flow was interpreted as resulting from equilibrium of the forces acting on the fluid flowing within a macroscopic sample of porous material. This line of reasoning can be extended to obtain a general expression for the law of flow in three dimensions in differential form.

Consider an element of volume in a field of fluid flow through a porous material. Let the element be of length δs and plane cross-sectional area δA. The vector δs, perpendicular to the cross section, defines the orientation of the element in the field of flow. Let the mean flow rate per unit area in the region of the element be $\hat{v}$. Then, a viscous force

$$F_\mu = -\mu B \delta A \hat{v} \cdot \widehat{\delta s} \tag{3-25}$$

acts parallel to δs on the fluid within the element of volume. Also acting parallel to δs is the net force due to the pressures on the ends of the element. This can be expressed as

$$F_p = -\phi \delta A \bigtriangledown p \cdot \widehat{\delta s} \tag{3-26}$$

The remaining force acting on the fluid within the element of volume is the weight of the fluid itself. Thus

$$F_g = -\rho g \phi \delta A \hat{1}_3 \cdot \widehat{\delta s} \tag{3-27}$$

is the component of this force parallel to $\widehat{\delta s}$. Here, $\hat{1}_3$ is a unit vector taken positive upward.

For steady flow, these forces must be in equilibrium, thus

$$-(\nabla p + \mu \frac{B}{\phi} \hat{v} + \hat{1}_3 \rho g) \cdot \widehat{\delta s} \phi \delta A = 0 \tag{3-28}$$

But since this must hold for every orientation of the volume element, it follows that

$$\nabla p + \mu \frac{B}{\phi} \hat{v} + \hat{1}_3 \rho g = 0 \tag{3-29}$$

or

$$\hat{v} = -\frac{\phi}{\mu B} (\nabla p + \hat{1}_3 \rho g) \tag{3-30}$$

Or with $K = \phi/B$ defined as before

$$\hat{v} = -\frac{K}{\mu} (\nabla p + \hat{1}_3 \rho g) = -\frac{K}{\mu} \left[\hat{1}_1 \frac{\partial p}{\partial x_1} + \hat{1}_2 \frac{\partial p}{\partial x_2} + \hat{1}_3 \left(\frac{\partial p}{\partial x_3} + \rho g \right) \right] \tag{3-31}$$

where $\hat{1}_1$, $\hat{1}_2$ and $\hat{1}_3$ are unit vectors parallel to the respective orthogonal Cartesian axes, x_1, x_2 and x_3. The coordinate, x_3, is measured positive upward. This equation is the logical generalization of the linear form given in section 3.11.

Note that in this equation all quantities, including K represent average values over the infinitesimal element of volume $\delta A \delta s$. This is, of course, a pure abstraction since K ceases to have physical meaning for infinitesimal samples of porous material. However, it is not this differential law which is compared to experiment for verification, but integrals of this law must agree with observation. Indeed, it is found that predictions based on this differential law of flow are in agreement with observation.

The differential law of flow for *incompressible* fluids can be expressed in very compact form by defining a flow potential as

$$\psi' = p + \rho g x_3 \tag{3-32}$$

Then, the law of flow becomes

$$\hat{v} = -\frac{K}{\mu} \nabla \psi' \tag{3-33}$$

Homogeneous Compressible Fluids. For homogeneous fluids under isothermal conditions the fluid density, ρ, is a single-valued function of the pressure, p, within the fluid. Consequently, a flow potential, ψ, can be defined as[13]

$$\psi = g x_3 + \int_{p_0}^{p} \frac{dp}{\rho(p)} \tag{3-34}$$

Then the differential form of Darcy's law (equation 3-31) is retained for compressible fluids as well as for incompressible fluids if the flux density, $\hat{v}$, is defined by

$$\hat{v} = -\frac{K\rho}{\mu} \nabla \psi \tag{3-35}$$

Thus, substituting in this equation the definition of ψ given above, there results

$$\hat{v} = -\frac{K}{\mu} (\nabla p + \hat{1}_3 \rho g) \tag{3-36}$$

which is the same form as given for incompressible fluids. The minus sign in equation (3-35) indicates that flow takes place in the direction of decreasing potential. Here $\hat{v}$ is measured at the local value of pressure.

3.31: Simultaneous Flow of Immiscible Fluids through Porous Media. Relative Permeability

Darcy's law for the flow of a homogeneous fluid through a porous medium can be extended readily to the simultaneous flow of two immiscible fluids by extending the concept of permeability and bringing in the concept of capillary pressure presented in section 2.21.

It is postulated that for the simultaneous flow of two immiscible fluids through a porous medium there exists an *effective permeability* and a flow potential for each fluid. These effective permeabilities, say K_w for the wetting fluid and K_{nw} for the non-wetting fluid, must each be less than or equal to the single-fluid permeability, K, of the medium. Therefore, relative permeabilities for the respective fluids can be defined as

$$k_w = \frac{K_w}{K} \leq 1 \tag{3-37}$$

and

$$k_{nw} = \frac{K_{nw}}{K} \leq 1 \tag{3-38}$$

where k_w and k_{nw} are dimensionless positive quantities.

The flow potentials for the two fluids are defined as

$$\psi_w = gx_3 + \int_{p_0}^{p_w} \frac{dp}{\rho_w(p)} \tag{3-39}$$

and

$$\psi_{nw} = gx_3 + \int_{p_0}^{p_{nw}} \frac{dp}{\rho_{nw}(p)} \tag{3-40}$$

Here p_0 is a reference pressure, p_w and p_{nw} are the pressures within the respective fluids and ρ_w and ρ_{nw} are the respective fluid densities.

Then for each phase, Darcy's law is applied in the form

$$\hat{v}_w = -k_w \frac{K\rho_w}{\mu_w} \nabla\psi_w \tag{3-41}$$

for the volume flux density of wetting fluid, and

$$\hat{v}_{nw} = -k_{nw} \frac{K\rho_{nw}}{\mu_{nw}} \nabla\psi_{nw} \tag{3-42}$$

for the volume flux density of non-wetting fluid. Here μ_w and μ_{nw} are the fluid viscosities of wetting and non-wetting fluids, respectively.

The fluid pressures, p_w and p_{nw}, are related by the capillary-pressure equation

$$p_c = p_{nw} - p_w \tag{3-43}$$

This, of course, constitutes an assumption. It is assumed, and this is borne out by experiment[2] that the capillary pressure-saturation relationship determined at static conditions also applies to the dynamic conditions of flow.

It is necessary to distinguish which fluid is displacing the other so that a choice between the imbition and drainage capillary-pressure curves can be made.

This concept and also that of capillary pressure can be extended to the simultaneous flow of three immiscible fluids, such as an oil-water-gas system. Thus, if the three fluids are denoted by the subscripts o, w, and g for oil, water and gas respectively, then relative permeabilities, k_o, k_w and k_g can be defined as was done for two fluids. Also, three flow potentials, ψ_o, ψ_w and ψ_g can be defined.

The pressures in the respective fluids cannot all be equal at any point because of interfacial tensions. Therefore, two capillary-pressure functions are defined by

$$p_o - p_w = p_{ow} \tag{3-44}$$

and

$$p_w = p_g = p_{wg} \tag{3-45}$$

Here p_o, p_w and p_g are the pressures in the respective fluids, p_{ow} is the capillary pressure between oil and water, and p_{wg} is the capillary pressure between water and gas. The capillary pressure between oil and gas is then given by

$$p_{og} = p_{ow} + p_{wg} \tag{3-46}$$

With these definitions, Darcy's law can be used to write an expression for the volume flux of each fluid.

3.32: Darcy's Law for Anisotropic Porous Media

In the foregoing discussion of the laws of fluid flow through porous media, it was tacitly assumed that permeability, and also relative permeabilities, are independent of the direction of fluid flow within the medium. This is not generally true for all porous media.

Many porous materials exhibit a distinct anisotropic character, particularly fibrous materials, such as wood as well as some sedimentary rocks. Thus, the fluid transmissibility in such materials is not the same in all directions. To take this characteristic of porous media into account requires a further generalization of the laws of flow. This extension is achieved by heuristic reasoning just as the previous extensions from the fundamental experimental law were made. The correctness of such extension can be established only by appeal to experiments for confirmation of predictions based on such extensions.

For isotropic porous media, Darcy's law presents simple proportionalities between the components of the volumetric flux and the corresponding components of the gradient of flow potential. Thus

$$v_i = -\frac{K\rho}{\mu} \frac{\partial \psi}{\partial x_i}, \qquad i = 1,2,3 \tag{3-47}$$

The most general linear relationship between the v_i and the components

$\partial\psi/\partial x_i$ that can be postulated takes the form

$$v_i = -\frac{\rho}{\mu}\left(K_{i1}\frac{\partial\psi}{\partial x_1} + K_{i2}\frac{\partial\psi}{\partial x_2} + K_{i3}\frac{\partial\psi}{\partial x_3}\right) \qquad i = 1,2,3 \qquad (3\text{-}48)$$

Of course, an additive constant could be included but this is physically impossible. Here the nine quantities, K_{ij}, $(i = 1,2,3; j = 1,2,3)$ form the elements of a tensor.

The generalization of Darcy's law postulated here retains a linear dependence of the v_i upon the components of potential gradient, which is a heuristic reason for assuming this form. This same form can also be arrived at by a basic set of postulates concerning the nature of the flow in anisotropic media.[9] However, our approach is to postulate a particular form and consider the consequences.

The three equations given in equation (3-48) can be written as the single matrix equation

$$\begin{pmatrix} v_1 \\ v_2 \\ v_3 \end{pmatrix} = -\frac{\rho}{\mu}\begin{pmatrix} K_{11} & K_{12} & K_{13} \\ K_{21} & K_{22} & K_{23} \\ K_{31} & K_{32} & K_{33} \end{pmatrix}\begin{pmatrix} \dfrac{\partial\psi}{\partial x_1} \\[2ex] \dfrac{\partial\psi}{\partial x_2} \\[2ex] \dfrac{\partial\psi}{\partial x_3} \end{pmatrix} \qquad (3\text{-}49)$$

Then, if a rotation of the corrdinate axis is considered, the manner in which the K-matrix transforms under such a rotation can be investigated. Such an investigation shows that, if the K-matrix is symmetric, then rotation of the axes to a particular orientation produces a diagonal K-matrix.

Thus, if

$$K_{ij} = K_{ji}, \qquad i = 1,2,3,; j = 1,2,3, \qquad (3\text{-}50)$$

then for a particular set of rectangular axis, x_i', $i = 1,2,3$ (i.e. a particular orientation of the coordinate system) the K-matrix takes the form (denoted as the K'-matrix)

$$(K'\text{-matrix}) = \begin{pmatrix} K_1 & 0 & 0 \\ 0 & K_2 & 0 \\ 0 & 0 & K_3 \end{pmatrix} \qquad (3\text{-}51)$$

The directions of the particular set of coordinate axis to which this K'-matrix corresponds are called the principal axes of the porous medium. Note that these principal axes are orthogonal to each other. Thus, the converse statement is also correct; for a porous medium having orthogonal principal axes the K-matrix is symmetric for any orientation of the coordinate system, and is diagonal for a coordinate system congruent with the principal axes.

For the coordinate axes oriented parallel to the principal axes of the porous medium having orthogonal principal axes, the postulated form of Darcy's law becomes

$$v_i = -K_i \frac{\rho}{\mu} \frac{\partial \psi}{\partial x_i}, \qquad\qquad i = 1,2,3 \qquad (3\text{-}52)$$

Thus, each component of ϑ is proportional to the corresponding component of the potential gradient but the constants of proportionality are not equal since the K_i are not all equal.

It should be noted that this rotation of axes also requires a change in the form of ψ. Thus, since, in general, not one of the primed coordinates is parallel to the vertical (direction of the gravitational force), ψ must be written as

$$\psi = g \sum_{i=1}^{3} x_i' \cos \alpha_i + \int_{p_0}^{p} \frac{dp}{\rho(p)} \qquad (3\text{-}53)$$

Here the α_i, $i = 1,2,3$, are the angles between the respective primed axes, x'_i, $i = 1,2,3$ and x_3 which was assumed vertical.

It has been experimentally demonstrated that some anisotropic porous media can be described by a permeability matrix as presented above.[17] Furthermore, the particular materials investigated possessed orthogonal principal axes. However, it cannot be expected that all anisotropic porous media would have orthogonal principal axes.

In the case of the simultaneous flow of immiscible fluids through anisotropic porous media, no experimental data have been reported to guide the extension of Darcy's law to such cases. However, the simplest and most reasonable postulate is that the relative permeabilities are independent of direction in the medium. Thus, for the coordinate axes oriented parallel to the principal axes of the medium only the diagonal K's need be considered. Therefore, postulate that the relative permabilities defined as

$$k_w = \frac{K_{1w}}{K_1} = \frac{K_{2w}}{K_2} = \frac{K_{3w}}{K_3} \qquad (3\text{-}54)$$

and

$$k_{nw} = \frac{K_{1nw}}{K_1} = \frac{K_{2nw}}{K_2} = \frac{K_{3nw}}{K_3} \qquad (3\text{-}55)$$

are functions of saturation only. Here the K_{iw} and K_{inw}, $i = 1,2,3$ denote the values of the permeabilities for the respective fluids, during simultaneous flow while the K_i, $i = 1,2,3$ are the principal values for a single fluid.

3.40: Equations of State for Fluids*

The flow potential, ψ, defined by equation (3-34), involves the fluid density, ρ, as a function of the pressure, p, within the fluid. Thus, in order to apply the law of flow to a particular problem the dependence of ρ on p must be determined. For the cases of most interest this dependence can be given in rather simple analytical form.

The trivial case of incompressibility is, of course, the most simple. For this case, the equation of state is simply

$$\frac{d\rho}{dp} = 0 \tag{3-56}$$

which gives

$$\rho = \text{constant} \tag{3-57}$$

The next most simple case is that corresponding to constant compressibility. Thus, since the compressibility of the fluid, c, is defined, for isothermal conditions, by

$$c = -\frac{1}{V_f}\frac{dV_f}{dp} \tag{3-58}$$

where V_f denotes fluid volume; and the mass of fluid is invariant, it follows that

$$c = \frac{1}{\rho}\frac{d\rho}{dp} \tag{3-59}$$

Hence, for c independent of p integration yields

$$\rho = \rho_0 e^{c(p-p_0)} \tag{3-60}$$

where ρ_0 is the value of ρ at the reference pressure, p_0.

This particular equation of state applies rather well to most liquids, though the presence of large quantities of dissolved gases causes deviations.

For isothermal variations in pressure the equation of state for an ideal gas is given by the Boyle-Mariotte law. Thus

$$pV = \frac{m}{M} RT \tag{3-61}$$

where V is the volume occupied by the mass, m, of gas of molecular weight

* Multicomponent fluids and phase equilibrium are considered in chapter X.

M, R is the gas constant and T is the absolute temperature. Since the fluid density is defined as m/V, it follows that

$$\rho = \frac{M}{RT}\, p \tag{3-62}$$

Observe that in this case

$$\frac{1}{\rho}\frac{d\rho}{dp} = \frac{1}{p} \tag{3-63}$$

so that the compressibility of an ideal gas is just the reciprocal of the absolute pressure of the gas.

For real gases deviations from the ideal gas law are taken into account by introducing the Z-factor. This factor is simply a function of p empirically determined from the definition

$$\rho = \frac{M}{RT}\frac{p}{Z(p)} \tag{3-64}$$

In this case, the dependence of ρ on p may be rather complex.

Other cases, such as an ideal gas under adiabatic conditions, can also be considered. However, the particular cases considered above constitute the most important cases and are the only ones to be treated specifically in the discussions to follow.

3.50: The Continuity Equation

In flow phenomena of any kind, fluid flow, the flow of electricity, or heat flow, one of the most useful mathematical tools is that obtained from a conservation principle. A conservation principle is simply the statement of the fact that some physical quantity is conserved; that is, neither created nor destroyed.

To formulate a general statement of the conservation of some physical quantity consider, in the field of flow, an element of volume in the form of a rectangular parallelopiped having sides Δx_1, Δx_2 and Δx_3. Let the concentration of the quantity be Γ, expressed in units per unit volume, and let the flux density of the quantity be denoted by $\hat{\Omega}$, expressed in units transported per unit time per unit area. Since $\hat{\Omega}$ is a vector quantity there are three independent components, Ω_1, Ω_2 and Ω_3 parallel to the respective coordinate axes. Furthermore, let the physical quantity in question be released within the field of flow at the rate G units per unit time per unit volume. Γ, $\hat{\Omega}$ and G are all variable quantities.

With the above definitions the conservation of the physical quantity can be mathematically stated. The average flux density across each face of the

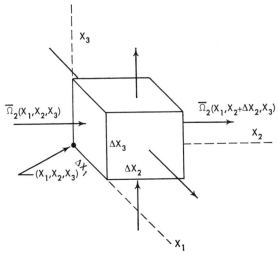

Figure 3-4. Volume element in region of fluid flow.

volume element can be indicated as shown in Figure 3-4. Here $\bar{\Omega}_1(x_1, x_2, x_3, t)$, for example, is the average value of Ω_1 over the face with sides Δx_2 and Δx_3 located at x_1; similarly, $\bar{\Omega}_1(x_1 + \Delta x_1, x_2, x_3, t)$ is the average value of Ω_1 over the corresponding face located at $x_1 + \Delta x_1$ and similarly for the other faces.

The amount of this physical quantity entering the volume element during the time interval t to $t + \Delta t$ is

$$[\bar{\Omega}_1(x_1, x_2, x_3, t)\Delta x_2 \Delta x_3 + \bar{\Omega}_2(x_1, x_2, x_3, t)\Delta x_1 \Delta x_3 + \bar{\Omega}_3(x_1, x_2, x_3, t)\Delta x_1 \Delta x_2]\Delta t$$

Similarly, the amount leaving the volume element during this time is

$$[\bar{\Omega}_1(x_1 + \Delta x_1, x_2, x_3, t)\Delta x_2 \Delta x_3 + \bar{\Omega}_2(x_1, x_2 + \Delta x_2, x_3, t)\Delta x_1 \Delta x_3$$
$$+ \bar{\Omega}_3(x_1, x_2, x_3 + \Delta x_3, t)\Delta x_1 \Delta x_2]\Delta t$$

Also, during this time, an amount

$$\bar{G}(x_1, x_2, x_3, t)\Delta x_1 \Delta x_2 \Delta x_3 \Delta t$$

is released within the element of volume. Here $\bar{G}$ indicates the average value of G within the element at time t.

As a result of the possible excess of inflow over outflow and the release of material within the element, the amount of the physical quantity within the element is increased during the time Δt by an amount

$$[\bar{\Gamma}(x_1, x_2, x_3, t + \Delta t) - \bar{\Gamma}(x_1, x_2, x_3, t)]\Delta x_1 \Delta x_2 \Delta x_3$$

where $\bar{\Gamma}$ denotes the average value of Γ within the volume element at the indicated time.

Since the physical quantity in question must be conserved, it follows that

(amount in) $-$ (amount out) $+$ (amount generated) $=$ (increase in content)

Therefore, substituting in this equation the expressions given above, dividing through by $\Delta x_1 \Delta x_2 \Delta x_3 \Delta t$ and taking the limits as Δx_1, Δx_2, Δx_3 and Δt are each allowed to approach zero, there results

$$-\left(\frac{\partial \Omega_1}{\partial x_1} + \frac{\partial \Omega_2}{\partial x_2} + \frac{\partial \Omega_3}{\partial x_3}\right) + G = \frac{\partial \Gamma}{\partial t} \tag{3-65}$$

This is the equation of continuity including a generation term, G. This can be written in more concise form by employing the divergence operator $(\nabla \cdot)$. Thus

$$-\Delta \cdot \hat{\Omega} + G = \frac{\partial \Gamma}{\partial t} \tag{3-66}$$

While the above formulation of the equation of continuity gives some insight into the physical meaning of this equation, it is rather clumsy. A more concise and rigorous derivation is as follows.

Since the physical quantity in question must be conserved, it follows that for an arbitrary element of volume

$$-\oint \hat{\Omega} \cdot \widehat{dA} + \iiint G dx_1 dx_2 dx_3 = \frac{d}{dt} \iiint \Gamma dx_1 dx_2 dx_3 \tag{3-67}$$

where the first integral is a surface integral taken over the closed surface of the volume element and the volume integrals are extended over the volume, V, of the element.

Since the volume element is considered fixed, the time derivative can be taken inside the integrals as a partial derivative operator on Γ. Then applying the divergence theorem to the surface integral and rearranging, there results

$$\iiint \left(-\nabla \cdot \hat{\Omega} + G - \frac{\partial \Gamma}{\partial t}\right) dx_1 dx_2 dx_3 = 0 \tag{3-68}$$

and since this must hold for any arbitrary volume element the integrand must be zero. Therefore, equation (3-66) follows.

Applications of the continuity equation to problems of fluid flow through porous materials are considered in the next section.

3.60: The Differential Equations of Fluid Flow through Porous Materials

Having formulated the mathematical statement of the law of flow through porous media and constructed the equation of continuity the differential equations describing fluid flow through porous media can be written. While it is possible to write these differential equations in a very general form it is more convenient to consider different types of flow independently. Furthermore, the differential equations formulated are not unique in form. The form selected will, in general, be different for different types of problems. Many of the equations formulated here are treated at greater length in later chapters.

Single-Phase Incompressible Flow. For incompressible fluids the volume of an element of fluid is not altered by changes in pressure. Hence, fluid volume is conserved and in the general equation of continuity $\hat{\Omega}$ becomes the volumetric flux density $\hat{v}$. Also, the concentration Γ becomes the concentration of fluid volume which is just the porosity, ϕ.

Thus, since ϕ is here considered constant, the equation of continuity in rectangular Cartesian coordinates becomes

$$\nabla \cdot \hat{v} = \frac{\partial v_1}{\partial x_1} + \frac{\partial v_2}{\partial x_2} + \frac{\partial v_3}{\partial x_3} = G(x_1, x_2, x_3, t) \tag{3-69}$$

Or, if, as is usually the case, there are no fluid sources or sinks within the region of flow, G is zero and this becomes $\nabla \cdot \hat{v} = 0$.

In this equation the components of flux density, v_1, v_2, v_3 are to be expressed in terms of the components of potential gradient by Darcy's law. Thus, several different equations result depending on the nature of the porous medium.

For an isotropic porous medium equation (3-35) apply for the v_i and the resulting differential equation is

$$\frac{\partial}{\partial x_1}\left(\frac{K\rho}{\mu}\frac{\partial \psi}{\partial x_1}\right) + \frac{\partial}{\partial x_2}\left(\frac{K\rho}{\mu}\frac{\partial \psi}{\partial x_2}\right) + \frac{\partial}{\partial x_3}\left(\frac{K\rho}{\mu}\frac{\partial \psi}{\partial x_3}\right) = 0 \tag{3-70}$$

Here the fluid viscosity, μ, is usually considered constant but the permeability, K, may, in general, be a function of the x_i.

If the medium is homogeneous then the permeability, K, is constant, and if μ is also constant the differential equation of flow takes the form of Laplace's equation. Thus

$$\nabla^2 \psi' = \frac{\partial^2 \psi'}{\partial x_1{}^2} + \frac{\partial^2 \psi'}{\partial x_2{}^2} + \frac{\partial^2 \psi'}{\partial x_3{}^2} = 0 \tag{3-71}$$

where the modified flow potential, ψ', is defined as

$$\psi' = p + \rho g x_3 \tag{3-72}$$

Since ρ and g are both constant, this can be written as

$$\frac{\partial^2 p}{\partial x_1{}^2} + \frac{\partial^2 p}{\partial x_2{}^2} + \frac{\partial^2 p}{\partial x_3{}^2} = 0 \tag{3-73}$$

Hence the only differences in the distributions of ψ' and p arise from the boundary conditions. These will be discussed elsewhere.

For the case of a homogeneous anisotropic porous medium, the components of volumetric flux density, v_i, $i = 1,2,3$ are given by equation (3-52) provided the coordinate axes are assumed oriented parallel to the principal axes of the porous medium. In this case, K_1, K_2 and K_3 are distinct constants and the differential equation of flow takes the form

$$K_1 \frac{\partial^2 \psi'}{\partial x_1{}^2} + K_2 \frac{\partial^2 \psi'}{\partial x_2{}^2} + K_3 \frac{\partial^2 \psi'}{\partial x_3{}^2} = 0 \tag{3-74}$$

Here the fluid viscosity, μ, has been taken as constant and hence could be eliminated by multiplying the equation through by μ.

Note that in this case the modified flow potential, ψ', is given by

$$\psi' = p + \rho g \sum_{i=1}^{3} x_i \cos \alpha_i$$

where the α_i, $i = 1,2,3$ are the angles between the respective axes and the vertical.

For this case a particular modification of the coordinate system permits an important simplification. Define the contracted coordinates, η_i, $i = 1,2,3$ as

$$\eta_i = x_i \sqrt{\frac{K_1}{K_i}}, \; i = 1,2,3 \tag{3-75}$$

Then, the differential equation of flow can be written as

$$\frac{\partial^2 \psi'}{\partial \eta_2{}^2} + \frac{\partial^2 \psi'}{\partial \eta_2{}^2} + \frac{\partial^2 \psi'}{\partial \eta_3{}^2} = 0 \tag{3-76}$$

with ψ' expressed in terms of the η_i as

$$\psi' = p(\eta_1, \eta_2, \eta_3, t) + \rho g \sum_{i=1}^{3} , \eta_i \sqrt{\frac{K_1}{K_i}} \cos \alpha_i \tag{3-77}$$

Thus both ψ' and p satisfy Laplace's equation in the η_i-coordinate system.

The differential equations given above are all expressed for the case of three-dimensional flow. Of course, if the geometry and boundary conditions insure either one- or two-dimensional flow, the equations can be modified by setting the proper derivatives equal to zero.

Flow of a Single-Phase Compressible Fluid. In the flow of a compressible fluid through a porous medium the volume of an element of fluid may change due to changes in pressure. Therefore, fluid volume is not conserved. However, the mass of an element of fluid remains unchanged during such variations in volume and hence mass is conserved.

The mass flux density now plays the part of $\hat{\Omega}$ in the equation of continuity and the fluid mass concentration plays the part of concentration. Thus

$$\hat{\Omega} = \rho \hat{v} \tag{3-78}$$

and

$$\Gamma = \phi \rho \tag{3-79}$$

where, as always, ϕ denotes fractional porosity and ρ denotes the fluid density.

Now the equation of continuity appears as

$$-\nabla \cdot (\rho \hat{v}) + G = \frac{\partial(\phi \rho)}{\partial t} \tag{3-80}$$

Since regions free of sources and sinks are of most general interest, G will be taken as zero in the following discussion.

Usually the porosity, ϕ, may be considered constant and hence can be placed outside the derivative operator.

In the case of an isotropic porous medium this becomes, with the v_i expressed in terms of the potential gradient components by Darcy's law

$$\nabla \cdot \left(\frac{K\rho^2}{\mu} \nabla \psi \right) = \phi \frac{\partial \rho}{\partial t} \tag{3-81}$$

Or, in terms of the pressure gradient,

$$\nabla \cdot \left[\frac{K\rho}{\mu} \left(\nabla p + \hat{1}_3 \, \rho g \right) \right] = \phi \frac{\partial \rho}{\partial t} \tag{3-82}$$

In the event that gravitational effects are negligible, as is usually the case, this can be written as

$$\nabla \cdot \left(\frac{K\rho}{\mu} \nabla p \right) = \phi \frac{\partial \rho}{\partial t} \tag{3-83}$$

Now several different cases arise as various equations of state for the fluid may apply. For an ideal liquid

$$c = \frac{1}{\rho} \frac{d\rho}{dp} = \text{constant} \tag{3-84}$$

and, therefore

$$\rho \nabla p = \frac{1}{c} \nabla \rho \tag{3-85}$$

Thus, if c is considered constant, the differential equation for the flow of an ideal compressible liquid through an isotropic incompressible medium is for negligible gravitational effects

$$\nabla \cdot (K \nabla \rho) = \phi \mu c \frac{\partial \rho}{\partial t} \tag{3-86}$$

If, in addition, the porous medium is homogeneous, then K is uniform and this becomes

$$\nabla^2 \rho = \frac{\phi \mu c}{K} \frac{\partial \rho}{\partial t} \tag{3-87}$$

or, in rectangular Cartesian coordinates

$$\frac{\partial^2 \rho}{\partial x_1^2} + \frac{\partial^2 \rho}{\partial x_2^2} + \frac{\partial^2 \rho}{\partial x_3^2} = \frac{\phi \mu c}{K} \frac{\partial \rho}{\partial t} \tag{3-88}$$

This equation is of exactly the same form as the Fourier equation of heat conduction.

For liquids of slight compressibility a further modification of this equation is sometimes employed. If

$$\sum_{i=1}^{3} \frac{\partial^2 p}{\partial x_i^2} \gg \frac{c \sum_{i=1}^{3} \left(\frac{\partial p}{\partial x_i}\right)^2}{1 + c(p - p_0)} \tag{3-89}$$

then equation (3-88) can be written in the approximate form

$$\frac{\partial^2 p}{\partial x_1^2} + \frac{\partial^2 p}{\partial x_2^2} + \frac{\partial^2 p}{\partial x_3^2} = \frac{\phi \mu c}{K} \frac{\partial p}{\partial t} \tag{3-90}$$

This can be demonstrated by noting that the equation of state can be written as

$$\rho = \rho_0 e^{c(p-p_0)} = \rho_0 \left[1 + c(p - p_0) + \frac{1}{2!} c^2(p - p_0)^2 + \cdots \right] \tag{3-91}$$

Then performing the required differentiation for substitution into equation (3-88) but retaining only those terms up through c^2, the above result is obtained.

In the case of flow of an ideal gas through a porous medium, gravity effects are negligible. Thus, upon employing the equation of state (3-62) in equation (3-86), there results

$$\nabla \cdot \left(\frac{Kp}{\mu} \nabla p \right) = \phi \frac{\partial p}{\partial t} \tag{3-92}$$

If both K and μ are considered constant, this can be written in the very simple form

$$\nabla^2 p^2 = \frac{\phi \mu}{Kp} \frac{\partial p^2}{\partial t} \tag{3-93}$$

which though non-linear is quite similar to the corresponding equation for the compressible liquid above.

The flow equation for a real gas or for gas flow under adiabatic conditions can be obtained in the same manner as the above equations were derived. Also, one may consider the Klinkenberg effect in which case K is made to depend upon the pressure.

Simultaneous Flow of Immiscible Fluids. For the simultaneous flow of immiscible fluids two independent conservation principles must be invoked. Since the conservation of mass includes the conservation of volume as a special case for incompressible fluids, the conservation of mass can be applied to each fluid.

Denoting the fluids by the subscripts w and nw for wetting and non-wetting fluids, respectively, and noting that for the wetting fluid, for example, the mass concentration is given by

$$\Gamma = \phi S_w \rho_w \tag{3-94}$$

the continuity equation is

$$-\nabla \cdot (\rho_w \hat{v}_w) = \phi \frac{\partial (S_w \rho_w)}{\partial t} \tag{3-95}$$

Here it is assumed that ϕ is constant and that no sources or sinks are present. Upon employing Darcy's law in the form of equation (3-41) this yields

$$\nabla \cdot \left(\frac{K_w \rho_w^2}{\mu_w} \nabla \psi_w \right) = \phi \frac{\partial (S_w \rho_w)}{\partial t} \tag{3-96}$$

Similarly, the equation applicable to the non-wetting phase is

$$\nabla \cdot \left(\frac{K_{nw} \rho_{nw}^2}{\mu_{nw}} \nabla \psi_{nw} \right) = \phi \frac{\partial (S_{nw} \rho_{nw})}{\partial t} \tag{3-97}$$

In these two equations, the proper equations of state relating ρ_w to p_w, and ρ_{nw} to p_{nw} must be introduced; also p_w and p_{nw} must be related by the proper capillary-pressure function.

These equations are quite general. Particular cases offer many interesting problems. Furthermore, extension to three immiscible fluids can be carried out exactly in the same manner. However, further consideration of these equations will be deferred to a later section.

3.70: Boundary Conditions*

In all cases the equations governing the flow of fluids through porous media are second-order partial differential equations. Thus, in order to arrive at a solution of such flow problems, it is necessary to specify boundary conditions for the dependent function or its derivatives. This function is either the pressure or a flow potential, although, in some cases, another function may be introduced.

In the case of compressible fluids, the equations also involve a time derivative and hence the initial distribution of pressure, fluid density or other dependent variable must also be specified.

When the simultaneous flow of immiscible fluids is considered there are two dependent variables to be obtained simultaneously. Hence boundary conditions on both must be given as well as initial conditions.

The mathematical form of such boundary and initial conditions for various physical conditions are described here.

The Closed Boundary. Since at a closed boundary the fluid velocity normal to the boundary is zero, Darcy's law gives

$$v_n = \hat{v} \cdot \hat{n} = -\frac{K\rho}{\mu} \nabla\psi \cdot \hat{n} = -\frac{K\rho}{\mu} \frac{\partial\psi}{\partial l_n} = 0 \tag{3-98}$$

or simply

$$\frac{\partial\psi}{\partial l_n} = 0 \tag{3-99}$$

Here $\hat{n}$ is a unit vector normal to the boundary and l_n is distance measured parallel to $\hat{n}$. This must be modified for anisotropic media.

Fluid Entry or Exit. On any section of boundary across which fluid enters or leaves the porous medium several different conditions may obtain. For a homogeneous fluid, entry or exit may be from or into a reservoir maintained at uniform pressure (or more generally at uniform potential,

---

* The necessary boundary and initial conditions are, of course, determinable by the method of characteristics; however, for the simpler cases at least, physical reasoning is adequate.

ψ). In this case, the boundary condition is simply, either

$$p(x_1, x_2, x_3, t) = \text{constant} \tag{3-100}$$

or

$$\psi(x_1, x_2, x_3, t) = \text{constant} \tag{3-101}$$

as the case may be. Still more generally, the pressure or potential may be a specified function of time and position on the boundary.

The boundary condition most often applying to boundaries across which flow occurs is the specification of the velocity normal to the boundary. Thus, most often

$$v_n = -\frac{K\rho}{\mu} \frac{\partial \psi}{\partial l_n}$$

is a specified function of position and time on the boundary.

Discontinuity in the Porous Medium. Very frequently flow problems must be considered in which a discontinuity in the permeability of the medium exists. The proper boundary conditions at such a boundary are arrived at from physical considerations. Suppose, for example, that the domain of flow is composed of two regions, 1 and 2, having different permeabilities, K_1 and K_2. Then, in the case of a homogeneous fluid, for example, the mathematical problem is that of finding two functions, $p_1(x_1, x_2, x_3, t)$ applying in region 1 and $p_2(x_1, x_2, x_3, t)$ applying in region 2. Obviously, since only one value of pressure may obtain at any point, one requirement is

$$p_1 = p_2 \tag{3-102}$$

on the boundary.

Also since what enters the boundary from one side must come out on the other side, the velocities normal to the boundary must be equal on the two sides

$$\rho \frac{K_1}{\mu} \frac{\partial \psi_1}{\partial l_n} = \rho \frac{K_2}{\mu} \frac{\partial \psi_2}{\partial l_n} \quad \text{(on the boundary)} \tag{3-103}$$

Here ψ_1 and ψ_2 are the flow potentials in the two regions, ρ is the fluid density at the pressure existing at the boundary and l_n is distance measured normal to the boundary.

Simultaneous Flow of Immiscible Fluids. Consideration of boundary conditions applying to the simultaneous flow of immiscible fluids is deferred to those sections in which particular problems are considered.

EXERCISES

1. A vertical column of sandstone with lateral surface sealed is filled with dilute salt water; permeable electrodes are placed over the ends. Write the equation for the electrical potential difference between electrodes which will prevent the water from draining from the column.
2. Show that a discontinuity in flow potential, ψ or ψ', always exists at a fluid-fluid interface separating regions of a porous medium containing different fluids.
3. Show that a discontinuity in the direction of fluid flow exists at a discontinuity in permeability, unless flow is everywhere normal to the discontinuity in medium.
4. Show that if the porous medium is itself compressible, the equation of continuity for mass flow has the form

$$-\nabla \cdot (\rho \hat{v}) + G = \phi \left(1 - \frac{c_\phi}{c} \right) \frac{\partial \rho}{\partial t}$$

where $c_\phi = -\phi^{-1} d\phi/dp$ is the pore compressibility and c is the compressibility of the fluid.

References

1. Barrer, R. M., "Diffusion in and through Solids," Cambridge, England, The University Press (1941).
2. Brown, J. W., *Trans AIME*, **192**, 67 (1951).
3. Chatenever, A., and Calhoun, J. C., Jr., *Trans. AIME*, **195**, 149 (1952).
4. Darcy, H., "Les fontainer publiques de la ville de Dijon," Dalmont, Paris (1856).
5. Ergun, S., and Orning, A. A., *Ind. Eng. Chem.*, **41**, 1179 (1949).
6. Estes, R. K., and Fulton, P. F., *Trans. AIME*, **207**, 338 (1956).
7. Fancher, G. H., and Lewis, J. A., *Ind. Eng. Chem.*, **25**, 1139 (1933).
8. Fatt, I., *Trans. AIME*, **207**, 144 (1956).
9. Ferrandon, J., *Génie Civil*, **125**, 24 (1948).
10. Forchheimer, P., *Z. Ver. deuts. Ing.*, **45**, 1782 (1901).
11. Geffen, T. M., Owens, W. W., Parish, and D. R., Morse, R. A., *Trans. AIME*, **192**, 99 (1951).
12. Glasstone, S., "Textbook of Physical Chemistry," D. Van Nostrand Co., New York (1946).
13. Hubbert, M. K., *Trans. AIME*, **207**, 222 (1956).
14. Jordan, J. K., McCardell, W. M., and Hocott, C. R., *Oil Gas J.*, **55**, No. 19, 98 (1957).
15. Klinkenberg, L. J., *Drilling Prod. Proc.*, **200** (1941).
16. Philip, J. R., and DeVries, D. A., *Trans. Am. Geophy. V.*, **38**, 222 (1957).
17. Richardson, J. G., Kerver, J. K., Hafford, J. A., and Osoba, J. S., *Trans. AIME*, **195**, 187 (1952).
18. Scheidegger, A. E., *Geofis. Pura. Appl.*, **28**, 75 (1954).
19. Woodside, W., and Cliffe, J. B., *Soil Sci.*, **87**, 75 (1956).

4. STEADY LAMINAR FLOW OF HOMOGENEOUS FLUIDS

4.10: The Steady-State Regime

The steady-state flow regime is characterized by invariance with time of all physical variables. Thus in the steady state the distributions of pressure and fluid velocity are independent of time.

The continuity equation reduces to

$$\nabla \cdot (\rho \hat{v}) = 0 \tag{4-1}$$

for steady flow, since all partial derivatives with respect to time are zero. Special cases of such steady flow are treated in this chapter.

Techniques for the solution of problems of steady flow of homogeneous fluids are rather well developed and many solutions of such problems are to be found in the literature. Consequently, no attempt is made here to present a complete survey of problem types or methods of solution. The basic formulation of such problems is given and some of the most useful analytical tools for solution of problems are described. A few examples are given here and some additional examples are to be found in Chapter 7.

4.20: Linear Flow and the Measurement of Permeability

For linear, horizontal flow of a homogeneous incompressible fluid through an isotropic porous medium the combination of Darcy's law and the equation of continuity yields the differential equation

$$\frac{\partial}{\partial x}\left(K \frac{\partial p}{\partial x} \right) = 0 \tag{4-2}$$

Here the fluid viscosity, considered as constant, has been eliminated.

In general the permeability, K, may be a function of position. If the sample of porous medium is in the form of a long thin cylinder, then at any cross section the permeability can be considered uniform. Then K and hence p are functions of x only.

Integrating once with respect to x yields

$$K \frac{\partial p}{\partial x} = \text{constant} \tag{4-3}$$

Then multiplying by $-A/\mu$, where A is the cross-sectional area, yields by Darcy's law

$$-\frac{KA}{\mu}\frac{\partial p}{\partial x} = q = \text{constant} \tag{4-4}$$

where q is the volumetric flow rate. Now rearranging gives

$$-\frac{dp}{dx} = \frac{q\mu}{A}\frac{1}{K(x)} \tag{4-5}$$

and integration yields

$$p(0) - p(x) = \frac{q\mu}{A}\int_0^x \frac{dx}{K(x)} \tag{4-6}$$

If then the length of the sample is L and

$$\Delta p = p(0) - p(L) \tag{4-7}$$

the above result can be written as

$$q = -\frac{\bar{K}}{\mu} A \frac{\Delta p}{L} \tag{4-8}$$

where

$$\frac{1}{\bar{K}} = \frac{1}{L}\int_0^L \frac{dx}{K(x)} \tag{4-9}$$

The quantity $\bar{K}$ is the harmonic mean permeability.

This shows that the permeability actually measured in a linear flow experiment is the harmonic mean permeability. If the sample is homogeneous, then $\bar{K}$ becomes simply the homogeneous permeability K.

If a gas is employed in the measurement of permeability, then account must be taken of both compressibility and gas-slip. For linear flow of an ideal gas the flow equation becomes

$$\frac{\partial}{\partial x}\left(\frac{K}{\mu} p \frac{\partial p}{\partial x}\right) = \phi \frac{\partial p}{\partial t} \tag{4-10}$$

In the steady state the pressures are stationary and independent of time, so that

$$\frac{\partial p}{\partial t} = 0. \tag{4-11}$$

Now the Klinkenberg equation relates K and p, thus

$$K = K_\infty \left(1 + \frac{b}{p}\right) \tag{4-12}$$

where K_∞ is to be considered a function of x. The equation of flow is then

$$\frac{d}{dx}\left[K_\infty p \left(1 + \frac{b}{p}\right)\frac{dp}{dx}\right] = 0 \qquad (4\text{-}13)$$

where again μ is considered as invariant and hence could be eliminated. This equation can be integrated once immediately, thus

$$K_\infty p \left(1 + \frac{b}{p}\right)\frac{dp}{dx} = \text{constant} = a \qquad (4\text{-}14)$$

Rearranging then gives

$$(p + b)\frac{dp}{dx} = \frac{a}{K_\infty(x)} \qquad (4\text{-}15)$$

and then integration yields

$$\frac{1}{2}[p^2(0) - p^2(L)] + b[p(0) - p(L)] = a\int_0^L \frac{dx}{K_\infty(x)} \qquad (4\text{-}16)$$

or

$$\bar{K}_\infty \bar{p}\left(1 + \frac{b}{p}\right)\frac{\Delta p}{L} = a \qquad (4\text{-}17)$$

Here

$$\bar{p} = \frac{1}{2}[p(0) + p(L)] \qquad (4\text{-}18)$$

$$\Delta p = p(0) - p(L) \qquad (4\text{-}19)$$

and

$$\frac{1}{\bar{K}_\infty} = \frac{1}{L}\int_0^L \frac{dx}{K_\infty(x)} \qquad (4\text{-}20)$$

Thus, from Darcy's law

$$a = \frac{\bar{q}\mu\bar{p}}{A} \qquad (4\text{-}21)$$

where q is the volume flow rate at the mean pressure $\bar{p}$ and A is the cross-sectional area. So again a harmonic mean value is obtained, this time for K_∞, however.

A fault of the above treatment lies in considering the constant b as independent of x. Strictly speaking, in most naturally porous materials not only K_∞ but b also is a function of position.

The two cases of linear flow considered above point up an important consideration in permeability measurements. In naturally porous materials, the permeability is a random function of the space coordinate and hence a very short specimen presents a statistically poor sample. On the other hand, long samples are difficult to handle as well as to prepare. Consequently, the best solution to good measurement of a mean value is to compute the harmonic mean of many measured permeabilities on short samples.

4.30: Horizontal Plane Flow in Homogeneous Media; Mathematical Formulation

Here a horizontal stratum of a uniform porous medium is considered. It is assumed that both the upper and lower boundaries are impermeable. It is also assumed that the boundary conditions on other boundaries are such that no vertical flow occurs.

If, in addition, the vertical axis is a principal axis of permeability, then the condition of no vertical flow is by Darcy's law

$$v_3 = -\frac{K_3 \rho}{\mu} \frac{\partial \psi}{\partial x_3} = 0 \tag{4-22}$$

where

$$\psi = \int_{p_0}^{p} \frac{dp}{\rho} + g x_3 \tag{4-23}$$

as defined in section 3.31. Thus ψ is a function of the coordinates x_1 and x_2 only. However, both p and ρ will depend on the vertical coordinate x_3. Thus from equation (4-23) above

$$\frac{\partial p}{\partial x_3} = -\rho g \tag{4-24}$$

Under these conditions the horizontal components of velocity are by Darcy's law

$$v_1 = -\frac{K_1}{\mu} \rho \frac{\partial \psi}{\partial x_1} = -\frac{K_1}{\mu} \frac{\partial p}{\partial x_1} \tag{4-25}$$

and

$$v_2 = -\frac{K_2}{\mu} \rho \frac{\partial \psi}{\partial x_2} = -\frac{K_2}{\mu} \frac{\partial p}{\partial x_2} \tag{4-26}$$

Then the continuity equation becomes

$$\frac{\partial}{\partial x_1} \left(\frac{K_1}{\mu} \rho \frac{\partial p}{\partial x_1} \right) + \frac{\partial}{\partial x_2} \left(\frac{K_2}{\mu} \rho \frac{\partial p}{\partial x_2} \right) = 0 \tag{4-27}$$

Two general cases can be considered, either the fluid is a liquid or a gas. If the fluid is a gas, then the Klinkenberg effect must be taken into account but gravity effects can be neglected. Thus, for an ideal gas of molecular weight, M, equation (4-24) becomes

$$\frac{\partial p}{\partial x_3} = -\frac{Mg}{RT} p \tag{4-28}$$

for isothermal conditions. Hence

$$p(x_1, x_2, x_3) = p(x_1, x_2, 0)e^{-(Mg/RT)x_3} \tag{4-29}$$

and since Mg/RT is small, it follows that for values of x_3 not too great

$$p(x_1, x_2, x_3) \approx p(x_1, x_2, 0) \tag{4-30}$$

This shows that for gases gravity can be neglected. Thus, g is put equal to zero for gases, but the Klinkenberg effect must be taken into account. This is done by assuming

$$K_1 = K_{1\infty}\left(1 + \frac{b}{p}\right), \qquad K_2 = K_{2\infty}\left(1 + \frac{b}{p}\right) \tag{4-31}$$

Strictly speaking, the constant b may not have exactly the same value along both principal axes but the error thus introduced is far outweighed by the mathematical simplification obtained. With this assumption and the density expressed in terms of the pressure, the equation of flow becomes

$$K_{1\infty}\frac{\partial}{\partial x_1}\left[\frac{M}{RT\mu}(p + b)\frac{\partial p}{\partial x_1}\right] + K_{2\infty}\frac{\partial}{\partial x_2}\left[\frac{M}{RT\mu}(p + b)\frac{\partial p}{\partial x_2}\right] = 0 \tag{4-32}$$

This can be reduced to a much simpler form by defining a function $U(p)$ as

$$U(p) = \frac{M\sqrt{K_{1\infty}K_{2\infty}}}{\mu RT}\int_0^p (p + b)\, dp = \frac{M\sqrt{K_{1\infty}K_{2\infty}}}{\mu RT}\frac{p(p + 2b)}{2} \tag{4-33}$$

Then equation (4-32) can be written as

$$\frac{\partial^2 U}{\partial x_1^2} + \frac{K_{2\infty}}{K_{1\infty}}\frac{\partial^2 U}{\partial x_2^2} = 0 \tag{4-34}$$

where μ is considered constant. Or defining the new rectangular coordinates

$$x = x_1, \qquad y = x_2\sqrt{\frac{K_{1\infty}}{K_{2\infty}}} \tag{4-35}$$

this can be written as

$$\frac{\partial^2 U}{\partial x^2} + \frac{\partial^2 U}{\partial y^2} = 0 \tag{4-36}$$

which is Laplace's equation for $U(x, y)$.

In the case of liquids, compressibility effects are usually negligible but gravity effects may be significant. Thus ρ is taken as constant and equation (4-24) yields

$$p(x_1, x_2, x_3) = p(x_1, x_2, 0) - \rho g x_3 \tag{4-37}$$

for incompressible liquids. Thus, a simple linear transformation exists between the pressure distributions in parallel horizontal planes.

For this case, the equation of flow can be written as

$$\frac{K_1\rho}{\mu}\frac{\partial^2 p}{\partial x_1^2} + \frac{K_2\rho}{\mu}\frac{\partial^2 p}{\partial x_2^2} = 0 \tag{4-38}$$

or, in the x, y coordinates (equation 4-35)

$$\frac{\partial^2 p}{\partial x^2} + \frac{\partial^2 p}{\partial y^2} = 0 \tag{4-39}$$

But p is a function of x_3 as well as x and y. For this reason, it is best to introduce the function

$$U' = \frac{\sqrt{K_1 K_2}}{\mu}(p + \rho g x_3)\rho \tag{4-40}$$

which, in view of equation (4-37), is independent of x_3. Thus

$$\frac{\partial^2 U'}{\partial x^2} + \frac{\partial^2 U'}{\partial y^2} = 0 \tag{4-41}$$

Hence for the steady horizontal flow of either an ideal isothermal gas or an incompressible liquid, there exists a potential function satisfying Laplace's equation in the x, y coordinate system. It is to be noted that if the medium is isotropic as well as homogeneous then the coordinates x and y are identical with the metric coordinates x_1 and x_2, respectively.

Since a wealth of mathematical techniques exist for the solution of Laplace's equation in two dimensions, the formulation of steady horizontal flow given here is most useful. These mathematical techniques are so important that a few of the more useful ones are described in the following sections. For a more complete treatment of mathematical methods, the reader is referred to treatises on heat conduction or potential theory.

4.31: The Stream Function and Complex Potential

Many problems of steady horizontal flow are most easily treated in terms of another function satisfying Laplace's equation. This function is called the stream function.

The stream function is intimately related to the potential function U, or

U', discussed in the previous section. The nature of this relation and the physical meaning of the stream function are established as follows.

Writing the ∇-operator in the x-y coordinate system as

$$\nabla \equiv \hat{\mathbf{i}}_x \frac{\partial}{\partial x} + \hat{\mathbf{i}}_y \frac{\partial}{\partial y} \qquad (4\text{-}42)$$

the differential equation for steady horizontal flow is (from section 4-30)

$$\nabla \cdot \nabla U = 0 \qquad (4\text{-}43)$$

Here U is to be considered as the function defined by equation (4-33), or equation (4-40), or some generalization thereof corresponding to a different equation of state. In any event, U is to be defined so that

$$-h \frac{\partial U}{\partial x} \, dy$$

represents rate of mass flow through the vertical strip of area hdx_2 in the x_1 direction and similarly

$$-h \frac{\partial U}{\partial y} \, dx$$

represents the rate of mass flow through the vertical strip of area hdx_1 in the x_2 direction.

In the case of the functions U and U' defined in the previous section, this condition is fulfilled.

Now an identity from vector calculus is

$$\nabla \cdot (\nabla \times \hat{V}) = 0 \qquad (4\text{-}44)$$

where $\hat{V}$ is any vector function. Thus, in view of equation (4-43) above, and the fact that only two dimensions are to be considered

$$\nabla U = \nabla \times \hat{V} \qquad (4\text{-}45)$$

and

$$\hat{V} = \hat{\mathbf{i}}_3 V(x, y) \qquad (4\text{-}46)$$

This yields the equations

$$\begin{cases} \dfrac{\partial U}{\partial x} = \dfrac{\partial V}{\partial y} \\[2mm] \dfrac{\partial U}{\partial y} = -\dfrac{\partial V}{\partial x} \end{cases} \qquad (4\text{-}47)$$

These equations are identical with the Cauchy-Riemann equations[1] of complex variable theory. This shows not only that a function $V(x, y)$ exists but also that the analytic function $Z(z)$ of the complex variable

$$z = x + iy, \quad (i = \sqrt{-1}) \tag{4-48}$$

defined by

$$Z(z) = U(x, y) + iV(x, y) \tag{4-49}$$

exists.

To see that $V(x, y)$ also satisfies Laplace's equation differentiate the first of equations (4-47) with respect to y, the second with respect to x and subtract the second result from the first. There results

$$\frac{\partial^2 V}{\partial x^2} + \frac{\partial^2 V}{\partial y^2} = 0 \tag{4-50}$$

Furthermore, in view of this result and Laplace's equation for U, it follows that

$$\frac{\partial^2 Z}{\partial x^2} + \frac{\partial^2 Z}{\partial y^2} = 0 \tag{4-51}$$

and the complex potential satisfies Laplace's equation.

The physical meaning of the stream function, $V(x, y)$, follows from equations (4-47) and the properties of the flow potential $U(x, y)$. Thus, the mass rate of flow through the vertical strip of area hdx_2 in the $-x_1$ direction is given by

$$h \frac{\partial V}{\partial y} \, dy$$

and the rate through the strip hdx_1 in the x_2 direction is given by

$$h \frac{\partial V}{\partial x} \, dx \, .$$

Consider the element of displacement,

$$\widehat{dl} = \hat{1}_x \, dx + \hat{1}_y \, dy,$$

in the x, y plane as shown in Figure 4-1. Since mass is conserved, it follows that for steady flow

$$\nabla V \cdot \widehat{dl}$$

is the rate of mass flow through the strip of area

$$h \sqrt{dx_1^2 + dx_2^2}$$

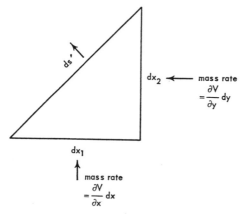

Figure 4-1. Geometry for interpretation of the stream function.

That is, across the line element

$$\widehat{dl'} = \hat{\imath}_1\, dx_1 + \hat{\imath}_2\, dx_2$$

as indicated in the figure.

This formulation of steady-plane flow shows that the field of flow can be described by two families of curves in the x, y plane, the two sets being orthogonal. The family of curves $V(x, y) = $ constant is everywhere parallel to the direction of fluid flow, while the family $U(x, y) = $ constant is everywhere perpendicular to the direction of flow. Note that these two families are not orthogonal in the $x_1 x_2$ plane unless the medium is isotropic.

It also follows from this formulation that every analytic function of the complex variable $z = x + iy$ represents the solution of a problem of steady-plane flow. The utility of this fact is illustrated by the examples presented in following sections.

In applying this formulation to flow problems the distinction between the x, y and x_1, x_2 planes must be borne in mind if the medium is anisotropic. In the x, y plane the lines of flow and the iso-potentials are always orthogonal but in the x_1, x_2 plane this will be the case only if the medium is isotropic. Furthermore, the transformation of boundary contours must be noted. In general, a circle in one plane is an ellipse in the other and vice versa. Similar remarks apply to other geometrical figures.

The proper procedure is to set up and solve the problem in the x, y plane and then to transform the results to the x_1, x_2 plane

It is to be observed that the solution of a problem of flow in an isotropic medium is also the solution of a problem of flow in an anisotropic medium of different geometry.

4.32: Radial Flow between Concentric Circular Cylinders

This problem illustrates the application of the mathematical formulations given in sections 4-30 and 4-31.

Frequently, the steady flow of oil into a well from the surrounding porous rock is represented as plane radial flow between concentric circular boundaries. The medium is considered *isotropic*. The interior circle represents the wall of the bore hole and the outer circle represents a boundary of constant potential called the "drainage radius."

A flow problem of this type can be formulated in several different ways. Thus, one may write Laplace's equation for the potential U in plane polar coordinates

$$\frac{1}{r}\frac{\partial}{\partial r}\left(r\frac{\partial U}{\partial r}\right) + \frac{1}{r^2}\frac{\partial^2 U}{\partial \theta^2} = 0, \qquad \begin{cases} r = \sqrt{x^2 + y^2} \\ \theta = \tan^{-1} y/x \end{cases} \quad (4\text{-}52)$$

and solve for U. Here it is observed that the symmetry of the problem, radial flow, requires that U be independent of θ and hence

$$\frac{1}{r}\frac{\partial}{\partial r}\left(r\frac{\partial U}{\partial r}\right) = 0 \tag{4-53}$$

is the equation to be solved.

Instead, one may write Laplace's equation for V in polar coordinates

$$\frac{1}{r}\frac{\partial}{\partial r}\left(r\frac{\partial V}{\partial r}\right) + \frac{1}{r^2}\frac{\partial^2 V}{\partial \theta^2} = 0 \tag{4-54}$$

and observe that the flow lines, lines of constant V, must be a family of straight lines radiating from the origin. Hence V must be independent of r and the equation to be solved is

$$\frac{\partial^2 V}{\partial \theta^2} = 0 \tag{4-55}$$

In lieu of either of these, one may write the complex variable $z = x + iy$ in polar form as

$$z = re^{i\theta} \tag{4-56}$$

and consider the complex potential

$$Z(re^{i\theta}) = U(r, \theta) + iV(r, \theta) \tag{4-57}$$

Since the flow is radial, the lines of constant U must be circles about the origin and the lines of constant V must be a radial family of straight lines. Thus

$$Z(re^{i\theta}) = U(r) + iV(\theta) \tag{4-58}$$

The simplest function $Z(z)$ yielding such a separation is

$$Z = A \ln z + B = A \ln (re^{i\theta}) + B \tag{4-59}$$

Thus, tentative solutions for U and V are

$$U = A \ln r + b \tag{4-60}$$

and

$$V = A\theta + b' \tag{4-61}$$

Here

$$B = b + ib' \tag{4-62}$$

The boundary conditions to be employed for $U(r)$ are

$$\begin{cases} U(r_w) = U_w \\ U(r_e) = U_e \end{cases} \tag{4-63}$$

Here r_w is the well radius and r_e is the drainage radius. Using these to evaluate A and b yields

$$U = \frac{U_e - U_w}{\ln r_e/r_w} \ln \frac{r}{r_w} + U_w \tag{4-64}$$

The boundary conditions on $V(\theta)$ follow from the fact that the total mass flow into the well must pass through an arc between $\theta = 0$ and $\theta = 2\pi$. Thus

$$V(0) = 0$$

$$V(2\pi) = -\frac{\rho_w q_w}{h} \tag{4-65}$$

are the boundary conditions on $V(\theta)$. Here ρ_w and q_w are the fluid density and volume flow rate at the well and h is the thickness of the stratum. Note that flow *into* the well is negative. These, when used to evaluate A and b', yield

$$V = -\frac{\rho_w q_w}{2\pi h} \theta \tag{4-66}$$

Furthermore, since the constant A was evaluated by two different procedures and the result must be the same

$$-\rho_w q_w = 2\pi h \frac{U_e - U_w}{\ln r_e/r_w} \tag{4-67}$$

The solutions for U and V thus obtained could have been obtained separately from the first procedures indicated. Also this last result, equation (4-67), could have been obtained directly from the solution 4-64 and Darcy's law, i.e.

$$\rho_w q_w = h \int_0^{2\pi} r_w \left(-\frac{\partial U}{\partial r} \right)_{r=r_w} d\theta \qquad (4\text{-}68)$$

Observe that for steady radial flow the mass flow rate is the same across all concentric circular cylinders.

4.33: Conformal Mapping

One of the most useful mathematical tools for the solution of steady-plane flow problems is conformal mapping of the complex plane. To understand the basis for this method of solution the properties of the complex potential, $Z(z)$, are considered.

The complex potential

$$Z(z) = U(x, y) + iV(x, y) \qquad (4\text{-}69)$$

generates the two families of curves, $U(x, y) =$ constant and $V(x, y) =$ constant. These sets of curves are orthogonal. Thus, in any infinitesimal region about a point z in the z plane a small rectangle can be constructed having curves $U =$ constant as two parallel sides and curves $V =$ constant as the other two parallel sides.

If a transformation of coordinates, from x, y to u, v is introduced, then one must consider the form which such an infinitesimal rectangle assumes in the $w = u + iv$ plane. If this small rectangle in the z plane also has a rectangular form in the w plane the transformation is said to be conformal. With respect to such transformations, the following theorem can be proved.[2]

"At each point where a function $f(z)$ is analytic and $df/dz \neq 0$, the mapping

$$w = f(z) \qquad (4\text{-}70)$$

is conformal." For example, the transformation

$$w = \ln z \qquad (4\text{-}71)$$

is conformal. Thus

$$u + iv = \ln re^{i\theta} = \ln r + i\theta \qquad (4\text{-}72)$$

or

$$\begin{cases} u = \ln (x^2 + y^2)^{1/2} \\ v = \tan^{-1} y/x \end{cases} \qquad (4\text{-}73)$$

is a conformal transformation.

Another important property of conformal transformations is that such transformations preserve the form of Laplace's equation. Thus suppose that $U(x, y)$ is a solution of

$$\frac{\partial^2 U}{\partial x^2} + \frac{\partial^2 U}{\partial y^2} = 0 \tag{4-74}$$

and a conformal transformation is invoked

$$z = f(w), \quad \begin{cases} x = x(u,v) \\ y = y(u,v) \end{cases} \tag{4-75}$$

Then when $U(x, y)$ is expressed in terms of u and v, $U[x(u, v), y(u, v)]$, it can be shown that

$$\frac{\partial^2 U}{\partial u^2} + \frac{\partial^2 U}{\partial v^2} = 0 \tag{4-76}$$

also. This can be stated in the form of a theorem as:[2] "Every harmonic function of x and y transforms into a harmonic function of u and v under the change of variables $x + iy = f(u + iv)$ where f is an analytic function." A harmonic function is a function satisfying Laplace's equation in the plane.

The manner in which these properties of conformal transformations, or mappings, facilitate the solution of fluid-flow problems is illustrated by several examples in later sections. The following example is designed to show how these techniques are applied.

Consider the problem of plane radial flow treated in the previous section. One can write the equation to be solved as

$$\frac{\partial^2 U}{\partial x^2} + \frac{\partial^2 U}{\partial y^2} = 0, \quad r_w^2 < x^2 + y^2 < r_e^2 \tag{4-77}$$

with the boundary conditions

$$\begin{cases} U = U_w \quad \text{on} \quad x^2 + y^2 = r_w^2 \\ U = U_e \quad \text{on} \quad x^2 + y^2 = r_e^2 \end{cases} \tag{4-78}$$

In the $z = x + iy$ plane the region of flow is bounded by two concentric circles about the origin. Observe the mapping of this region under the logarithmic transformation given by equation (4-71) or (4-73). This is illustrated in Figure 4-2.

The region of flow in the w plane is a rectangle and the equation to be solved is equation (4-76). The solution of the flow problem in this plane can be written down immediately since the flow regime is obviously just

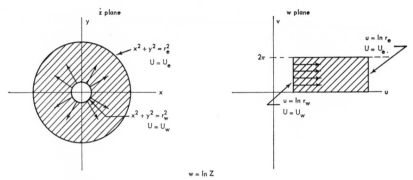

Figure 4-2. Illustrating the mapping of a radial flow geometry into a linear form.

linear flow. Thus

$$U = A + Bu \tag{4-79}$$

where A and B are constants. Employing the boundary conditions

$$\begin{cases} U = U_w & \text{on} \quad u = \ln r_w \\ U = U_e & \text{on} \quad u = \ln r_e \end{cases} \tag{4-80}$$

there results

$$U = U_w + (U_e - U_w) \frac{u - \ln r_w}{\ln r_e - \ln r_w} \tag{4-81}$$

But from (4-73) this is

$$U = U_w + \frac{U_e - U_w}{\ln r_e/r_w} \ln \frac{\sqrt{x^2 + y^2}}{r_w} \tag{4-82}$$

which is the solution of the problem in terms of x and y.

The procedure employed above is quite general. One simply constructs a conformal mapping $w = f(z)$ which maps the region of flow into some simple geometric form for which the solution is known, and then expresses this solution in terms of the original variables. The construction of the proper transformation is sometimes difficult, however, and there is no substitute for experience in such problems.

Observe that while the gross geometry of the above problem is radically altered by the transformation the microscopic features are preserved. Thus, in the original domain the curves $U = $ constant are circles orthogonal to the lines $V = $ constant, while in the transformed domain these same curves are the straight lines $U = $ constant which are orthogonal to the lines $V = $ constant.

4.34: The Schwarz-Christoffel Transformation; Conformal Mapping of a Rectangle

In the previous section the utility of conformal mapping in the solution of problems of steady plane flow was pointed out. However, the discovery of the proper mapping function for a given problem is not at all straightforward. Fortunately, there exists a procedure for constructing a mapping function which transforms the interior of a polygon in the z plane into the upper half of the w plane, the boundary of the polygon becomes the real axis in the w plane. This transformation is the Schwarz-Christoffel transformation.

Thus, suppose that the region of flow in the z plane is some figure bounded by straight-line segments, some of which are iso-potentials, others are streamlines. By applying the Schwarz-Christoffel transformation, this region is mapped into the upper half of the w plane. Now another Schwarz-Christoffel transformation is constructed which takes a rectangle from the η plane, say, into the w plane. The inverse of this transformation then takes the region of flow from the w plane to the interior of a rectangle in the η plane. Usually, this can be accomplished so that two parallel sides of the rectangle are streamlines and the other two sides are isopotentials. Thus, in this domain, linear flow obtains and the solution can be written down immediately.

Without digressing into the mathematical details of the formulation of the Schwarz-Christoffel transformation, a description of this transformation will be given.

Consider a polygon $ABCD$ in the z plane having external angles α, β, γ, δ as indicated in Figure 4-3. Note that

$$\alpha + \beta + \gamma + \delta = 2\pi \tag{4-83}$$

The interior of this polygon is to be mapped on the upper half of the w plane by

$$w = f(z) \tag{4-84}$$

the boundary of the polygon going onto the real axis, u. Thus, the vertices, A, B, C, D map into the points $w = a, b, c, d$ on the real axis as indicated in Figure 4-3. The transformation which accomplishes this must satisfy the equation

$$\frac{dz}{dw} = A(w - a)^{-(\alpha/\pi)}(w - b)^{-(\beta/\pi)}(w - c)^{-(\gamma/\pi)}(w - d)^{-(\delta/\pi)} \tag{4-85}$$

Here the constant A, which may be complex, determines the size and orientation of the polygon in the z plane.

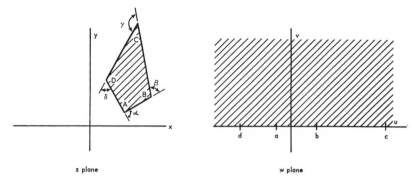

Figure 4–3. Mapping of a polygon on the upper half of the complex plane.

In general, equation (4-85) will have as many factors on the right, exclusive of the constant A, as there are vertices in the polygon. The exception occurs when a vertex is mapped into $w = \pm\infty$. In this case, the factor corresponding to this vertex will be omitted.

Equation (4-85) leads to rather complicated integrals for most cases of interest. Since the rectangle is the most useful case, it will be described in some detail.

To construct the Schwarz-Christoffel transformation for the rectangle we begin in the w plane. Let the images of two vertices, B, C, be located at $w = -1$ and $w = 1$, and the other two at $-1/\kappa$ and $1/\kappa$, respectively. For the rectangle, the external angles all have the value $\pi/2$, so equation (4-85) takes the form

$$\frac{dz}{dw} = \frac{A\kappa}{(1 - w^2)^{1/2}(1 - \kappa^2 w^2)^{1/2}} \tag{4-86}$$

Then if the origin in the z plane coincides with $w = 0$, this yields

$$z = \bar{A} \int_0^w \frac{dw}{(1 - w^2)^{1/2}(1 - \kappa^2 w^2)^{1/2}} \tag{4-87}$$

where $A\kappa$ is replaced by $\bar{A}$. But since $\bar{A}$ now only affects the scale in the z plane, it will be put equal to unity, thus

$$z = \int_0^w \frac{dw}{(1 - w^2)^{1/2}(1 - \kappa^2 w^2)^{1/2}} \tag{4-88}$$

Evaluating this integral for $w = 1$ yields

$$z(w = 1) = \mathbf{K}(\kappa) \tag{4-89}$$

and for $w = -1$ it yields

$$z(w = -1) = -\mathbf{K}(\kappa) \tag{4-90}$$

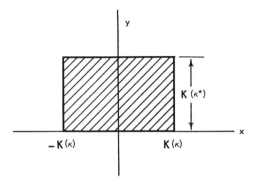

Figure 4-4. Schwarz-Christoffel mapping of the rectangle.

where $\mathbf{K}(\kappa)$ is the complete elliptic integral of the first kind with modulus κ. Numerical tables of $\mathbf{K}(\kappa)$ are available[3]. This shows that the width of the rectangle in the z plane is $2\mathbf{K}(\kappa)$.

Similarly, it can be shown that for $w = 1/\kappa$

$$z(w = 1/\kappa) = \mathbf{K}(\kappa) + i\mathbf{K}(\kappa^*) \tag{4-91}$$

and similarly for $w = -1/\kappa$

$$z(w = -1/\kappa) = -\mathbf{K}(\kappa) + i\mathbf{K}(\kappa^*) \tag{4-92}$$

where

$$\kappa^* = (1 - \kappa^2)^{1/2} \tag{4-93}$$

is the complementary modulus. Thus the rectangle in the z plane has height equal to $\mathbf{K}(\kappa^*)$ as indicated in Figure 4-4.

In view of the availability of tables of complete elliptic integrals, these expressions for the width and height of the rectangle are very useful, but for interior points in the rectangle, equation (4-88) represents an elliptic function and is rather difficult to employ. Even so, this transformation is extremely useful as is shown in the following section.

For a more complete discussion of conformal mapping and the Schwarz-Christoffel transformation, the reader may refer to Gibbs[2] or Churchill.[1]

4.40: Transverse Permeabilities of Cylindrical Cores

The problem treated here is one of practical importance in the petroleum industry and is presented to illustrate the utility of conformal mapping. In particular, the Schwarz-Christoffel transformation for the rectangle is employed.

Cores, or samples of reservoir rocks, are obtained from oil wells during drilling in the form of right circular cylinders with axis generally perpendicular to the bedding planes of the rocks.

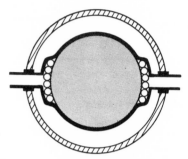

Figure 4–5. Plan view of transverse permeability apparatus.

Usually, the permeability of a sedimentary rock is not the same parallel and perpendicular to the bedding planes. Since the flow of oil to the well bore is generally parallel to the bedding planes, it is the permeability in this direction which is of most importance.

Thus, one may cut small circular plugs transverse to the axis of the core sample and measure the permeabilities of these plugs by linear-flow experiments, or one may attempt to employ a long section of the core itself and employ transverse flow. In the oil industry, the first course is usually followed but frequently the second course is called for. Thus, if the rock is a vugular limestone, small plugs are far from representative of the core.

An arrangement for producing transverse flow through a right circular cylinder of porous rock having plane ends is as follows. Rows of small coil springs are placed very close together in two groups parallel to the axis of the cylinder and running the full length of the cylinder. This is shown in the plan view in Figure 4-5.

Fitted around the cylinder and the groups of coil springs is a rubber sleeve. Thick circular disks of rubber are placed on the ends of the cylinder in the sleeve. The whole is then mounted within a large metal cylinder and screw plugs fitted into the ends so as to compress the rubber disks against the core, thus achieving an end seal.

Connected to the rubber sleeve at the center over each group of coil springs is a rigid tube leading through a seal out of the metal cylinder. The fluid to be used in the flow experiment enters the core through one of these and leaves through the other.

The annular space between the rubber sleeve and the metal cylinder is pressured up with compressed air to a pressure considerably in excess of the pressure in the core. This seals the rubber sleeve to the core thus preventing bypass flow around the core.

With this arrangement steady-plane flow of a gas transverse to the axis of the core can be achieved. For such flow, we have from section 4.30

$$\frac{\partial^2 U}{\partial x^2} + \frac{\partial^2 U}{\partial y^2} = 0 \qquad (4\text{-}94)$$

with the boundary conditions shown in Figure 4-6. Here only one-half of the circle is considered because of the symmetry in the flow geometry. The diameter connecting the midpoints of the entrance and exit sections is a streamline. This symmetry condition constitutes the basis for the boundary conditions in the figure. Note that since there is no flow across any streamline, a streamline is equivalent to a sealed boundary and hence the derivative of U normal to the streamline (or boundary) is zero.

To obtain the solution to this boundary-value problem, we first apply the conformal transformation

$$w = -i \ln \frac{z}{R} \qquad (4\text{-}95)$$

where R is the radius of the core. This yields the region of flow as a semi-infinite strip in the w plane as shown in Figure 4-7. Then we apply a second conformal mapping

$$w' = \frac{\sin w}{\kappa} \qquad (4\text{-}96)$$

which transforms the region of flow to the whole upper half of the w' plane. This is shown in Figure 4-8. Here

$$\kappa = \sin \left(\frac{\pi}{2} - \alpha \right) = \cos \alpha \qquad (4\text{-}97)$$

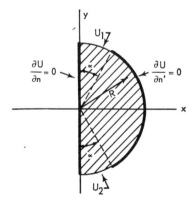

z plane

Figure 4-6. Boundary conditions for the transverse permeability problem.

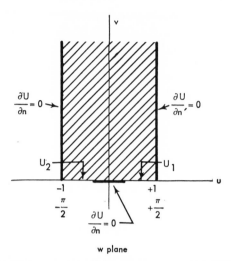

Figure 4–7. First mapping of the transverse permeability problem.

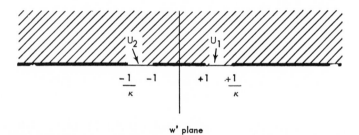

w' plane

Figure 4–8. Second mapping of the transverse permeability problem.

We can now apply the Schwarz-Christoffel transformation. In the previous section, it was shown that a rectangle of width $2\mathbf{K}(\kappa)$ and height $\mathbf{K}(\kappa^*)$ maps on the upper half of the image plane with the vertices going into the points $-\kappa^{-1}$, -1, $+1$, $+\kappa^{-1}$. Thus the inverse of the transformation

$$w' = \int_0^{w''} \frac{d\zeta}{(1 - \zeta^2)^{1/2}(1 - \kappa^2\zeta^2)^{1/2}} \tag{4-98}$$

transforms our domain of flow from the upper half of the w' plane into the rectangle in the w'' plane as shown in Figure 4-9. Here $\mathbf{K}'$ denotes $\mathbf{K}(\kappa^*)$.

Now we can write the equation for the flow rate through the core in terms of the geometry and flow-potential difference.

Letting the length of the core be L, the mass flow rate be $\dot{m}$ and noting that only one-half of this flow passes through the domain employed, we have

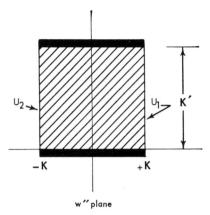

Figure 4–9. Final mapping of the transverse permeability problem.

$$\dot{m} = 2L \frac{\mathbf{K'}}{2\mathbf{K}} (U_1 - U_2) \qquad (4\text{-}99)$$

In particular, since an ideal gas is used and the Klinkenberg effect must be considered, U is represented by equation (4-33) of section 4.30. However, the core is considered isotropic in the plane, so $K_{1\infty} = K_{2\infty} = K_\infty$. Also we can write $\dot{m}$ as

$$\dot{m} = \frac{M}{RT} p_1 q_1 \qquad (4\text{-}100)$$

where p_1 is upstream pressure (entry) and q_1 is the volumetric flow rate of gas measured at this pressure. Thus

$$p_1 q_1 = \frac{K_\infty}{\mu} L \left(\frac{\mathbf{K'}}{\mathbf{K}} \right) [p_1(p_1 + 2b) - p_2(p_2 + 2b)] \qquad (4\text{-}101)$$

or

$$K_\infty = \frac{p_1 q_1 \mu}{2\bar{p} L \left(1 + \dfrac{b}{\bar{p}} \right) \Delta p} \left(\frac{\mathbf{K}}{\mathbf{K'}} \right) \qquad (4\text{-}102)$$

where

$$\bar{p} = \frac{1}{2} (p_1 + p_2) \qquad (4\text{-}103)$$

and

$$\Delta p = p_1 - p_2 \qquad (4\text{-}104)$$

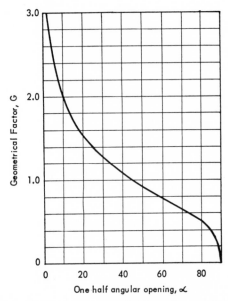

Figure 4–10. Geometric factor versus one-half angular opening for transverse permeability calculation.

Tables of elliptic integrals have been employed to plot the geometric factor $\mathbf{K}/\mathbf{K}'$ versus α. This is shown in Figure 4-10. Note that for an angular opening of $\pi/2$ for entry and exit, this factor is unity.

Thus, the technique of conformal mapping has yielded a formula for computing the transverse permeability of a circular core from measurable quantities.

4.50: Sources and Sinks

Quite frequently in problems of steady flow, the fluid is introduced into the porous medium over a very small region, or the fluid leaves the region of flow through such a small region. For example, in the waterflood technique of secondary recovery employed in the petroleum industry water is injected into the plane horizontal oil stratum at selected wells and oil and water are produced at other wells. From the standpoint of plane horizontal flow, the injection wells are circular sources of small radius, and the production wells are circular sinks of small radius.

In many problems of steady flow such sources and sinks can be treated as mathematical points. The distribution of flow potential due to such a source is deduced as follows.

We first consider steady incompressible flow in an infinite three-dimensional region and assume a flow potential ψ' to exist.

$$\hat{v} = -\frac{K}{\mu} \nabla \psi' \tag{4-105}$$

is the volume flux density, and

$$\frac{\partial^2 \psi'}{\partial x_1^2} + \frac{\partial^2 \psi'}{\partial x_2^2} + \frac{\partial^2 \psi'}{\partial x_3^2} = 0 \tag{4-106}$$

Further, it is assumed that at the point a, b, c fluid is introduced at a volumetric rate q. Since the flow has spherical symmetry about this point, the spherical coordinates

$$\begin{cases} r_s = [(x_1 - a)^2 + (x_2 - b)^2 + (x_3 - c)^2]^{1/2} \\[2mm] \theta = \cos^{-1} \dfrac{x_3 - c}{r_s} \\[2mm] \alpha = \cos^{-1} \dfrac{x_1 - a}{r_s \sin \theta} \end{cases} \tag{4-107}$$

are introduced.

Due to the spherical symmetry ψ' is a function of r_s only and Laplace's equation is

$$\frac{1}{r_s^2} \frac{\partial}{\partial r_s} \left[r_s^2 \frac{\partial \psi'}{\partial r_s} \right] = 0 \tag{4-108}$$

A solution of this equation is

$$\psi' = \frac{A}{r_s} + B \tag{4-109}$$

For this solution, the volume flux density is

$$\hat{v} = \hat{1}_r \frac{KA}{\mu r_s^2} \tag{4-110}$$

where $\hat{1}_r$ is a unit vector in the r_s direction. Then, since the total rate of flow through any spherical surface about the source point is q, it follows that:*

$$q = \oint r^2 v \, d\beta = 4\pi \frac{KA}{\mu} \tag{4-111}$$

or

* Here $d\beta$ denotes the element of solid angle.

$$A = \frac{q\mu}{4\pi K} \tag{4-112}$$

Thus, taking as a second-boundary condition

$$\lim_{r_s \to \infty} \psi' = \psi_\infty' = \text{constant} \tag{4-113}$$

we have

$$\psi' = \frac{q\mu}{4\pi K} \frac{1}{r_s} + \psi_\infty' \tag{4-114}$$

as the representation of a point source.

It should be noted that this solution also represents the potential distribution for a spherical source of any radius. Consequently, the point source can be considered as a limiting case of the spherical source. Also observe that if the point a, b, c is a sink then q is simply a negative quantity.

The property which makes point sources of such great utility in the solution of fluid-flow problems is that these solutions of the flow equation are subject to a superposition principle. This is shown as follows.

Consider two-point sources of strengths q and q' located at points a, b, c and a', b', c', respectively. If only the source q is present, the potential distribution is given by

$$\psi' = \frac{q\mu}{4\pi K} [x_1 - a)^2 + (x_2 - b)^2 + (x_3 - c)^2]^{-1/2} + \psi_\infty' \tag{4-115}$$

while if only q' is present a similar expression applies with q, a, b and c replaced by q', a', b' and c', respectively. We now show that, if both point sources are present, the potential distribution is given by

$$\psi' = \frac{q\mu}{4\pi K} [(x_1 - a)^2 + (x_2 - b)^2 + (x_3 - c)^2]^{-1/2}$$

$$\tag{4-116}$$

$$+ \frac{q'\mu}{4\pi K} [(x_1 - a')^2 + (x_2 - b')^2 + (x_3 - c')^2]^{-1/2} + \psi_\infty'$$

When this is substituted into equation (4-106) it is found that the differential equation for ψ' is satisfied. Also it is noted that the boundary condition at infinity, equation (4-113), is satisfied.

That this solution represents the two-point sources is verified by first writing equation (4-116) in the form

$$\psi' = \frac{q\mu}{4\pi K} [r_s'^2 + d^2 - 2r_s' d \cos \gamma]^{-1/2} + \frac{q'\mu}{4\pi K} \frac{1}{r_s'} + \psi_\infty' \tag{4-117}$$

where d is the distance between the two-point sources, and γ is the angle

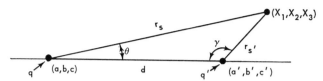

Figure 4-11. Geometry of two-point sources.

indicated in Figure 4-11. Then the flux out from the point source q' is given by

$$-\frac{K}{\mu}\frac{\partial \psi'}{\partial r_s'} = \frac{q}{4\pi}\frac{-\cos(\theta + \gamma)}{[r_s'^2 + d^2 - 2r_s'\,d\cos\gamma]} + \frac{q'}{4\pi r_s'^2} \qquad (4\text{-}118)$$

Now if this is integrated over the surface of a small sphere of radius r_s' about the source q', the total flux will be obtained. However, if

$$r_s' \ll d \qquad (4\text{-}119)$$

then equation (4-118) can be approximated by

$$-\frac{K}{\mu}\frac{\partial \psi'}{\partial r_s'} \approx \frac{q}{4\pi}\frac{(-2\cos\gamma)}{d^2} + \frac{q'}{4\pi r_s'^2} \qquad (4\text{-}120)$$

Then integration over the small sphere of radius r_s yields

$$q' = \int -\frac{K}{\mu}\left(\frac{\partial \psi'}{\partial r_s'}\right)_{r_s'} da \qquad (4\text{-}121)$$

where da is the element of area on the sphere. The approximation for small r_s' need not be employed in order to obtain this result but it simplifies the integration. Thus the solution represented by equation (4-116) yields the proper flow rate for the point source. A similar calculation yields a corresponding expression for the other point source.

This shows that the contributions of the point sources to the potential distribution are additive. Hence the potential distribution corresponding to any number of point sources can be written simply as a sum of terms of the form corresponding to a single-point source.

In two-dimensional flow the potential is a function of x_1 and x_2 only. For this case, a point source in the plane is a line source in three dimensions that is normal to the plane of flow.

The line source can be obtained as a superposition of point sources or by a direct procedure as follows.

Suppose fluid to be introduced at a constant rate q/h per unit length along a vertical line located at a, b in the horizontal x_1, x_2 plane. The potential ψ' must satisfy Laplace's equation in x_1, x_2 if the medium is isotropic.

An elementary solution of Laplace's equation in x_1 and x_2 is

$$\psi' = A \ln [(x_1 - a)^2 + (x_2 - b)^2]^{1/2} + C \tag{4-122}$$

This potential distribution yields the proper flow rate from the line if

$$A = \frac{q\mu}{2\pi Kh} . \tag{4-123}$$

Thus

$$\psi' = \frac{q\mu}{4\pi Kh} \ln [(x_1 - a)^2 + (x_2 - b)^2] + c \tag{4-124}$$

represents an isolated point sink in the plane except for the arbitrary constant c. This function is unbounded as x_1 or x_2 approaches infinity and hence does not represent a physically realizable potential distribution. Even so, this does adequately represent a realistic distribution of velocity in the neighborhood of a well.

Note that both the three-dimensional point source and the plane point source are represented by functions which are singular at the source. That is, the potential function is infinite at the source. It is just this property which permits the superposition of such sources because the flow distribution at the source is unaffected by other sources.

In the foregoing considerations an isotropic medium was assumed. This is not necessary. The same type of analysis can be carried out for the anisotropic case. Thus, for example, a plane point source in an anisotropic medium is represented by

$$\psi' = \frac{q\mu}{4\pi h \sqrt{K_1 K_2}} \ln \left[(x_1 - a)^2 + \frac{K_1}{K_2} (x_2 - b)^2 \right] + c \tag{4-125}$$

where it is assumed that the coordinate axes coincide with the principal axes of permeability.

Also note that it is not necessary that the fluid be incompressible in order to construct a point sink. For example, in steady plane flow the potential U, equation (4-33), may be employed and a point source constructed for the flow of an ideal gas. Point sources can be defined also for the transient flow of homogeneous compressible fluids.

4.51: System of Wells Near a Plane Discontinuity in Medium: The Method of Images

The superposition principle for point sources offers a method of solution to problems involving sources in inhomogeneous media. The problem of a system of oil wells in a plane horizontal stratum composed of two regions of different permeability meeting at a vertical plane can be solved.

To solve a problem of this type the method of images is employed. This procedure is formulated here for point sources in the plane but the method is quite general and can be formulated for point sources in three dimensions equally well.

We consider a horizontal isotropic stratum of uniform thickness and infinite in areal extent. This stratum is composed of two parts separated by a vertical plane, one part having permeability K_a and the other having permeability K_b.

The stratum is supposed filled with an incompressible fluid of density ρ and viscosity μ. A well, here to be represented by a point source, is supposed located in medium (a) at a distance d from the discontinuity.

The mathematical problem to be solved can be formulated by taking the origin of coordinates at a point on the discontinuity, the surface of discontinuity coinciding with the x_1 axis. Thus

$$\begin{cases} \dfrac{\partial^2 \psi_a'}{\partial x_1^2} + \dfrac{\partial^2 \psi_a'}{\partial x_2^2} = 0, & x_2 < 0 \\[2ex] \dfrac{\partial^2 \psi_b'}{\partial x_1^2} + \dfrac{\partial^2 \psi_b'}{\partial x_2^2} = 0, & x_2 > 0 \\[2ex] \psi_a' = \psi_b', & x_2 = 0 \\[2ex] K_a \dfrac{\partial \psi_a'}{\partial x_2} = K_b \dfrac{\partial \psi_b'}{\partial x_2}, & x_2 = 0 \end{cases} \tag{4-126}$$

Here $\psi_a'(x_1, x_2)$ is the potential function in medium (a) with permeability K_a and $\psi_b'(x_1, x_2)$ is the potential function in medium (b) of permeability K_b. In addition to the above requirements, which are very general, we fix a point source of strength q at $x_2 = -d$, $x_1 = 0$. Now the function

$$\psi_a' = \frac{q\mu}{4\pi K_a h} \ln [x_1^2 + (x_2 + d)^2] \tag{4-127}$$

satisfies the point-source requirement but does not satisfy the boundary condition on flow at the discontinuity. As a means of correcting this, we place another point source of strength Aq at $x_2 = +d$, $x_1 = 0$ in medium (b). This is the "image" of the point source at $x_2 = -d$, $x_1 = 0$. Here A is a constant to be determined.

Assume

$$\psi_a' = \frac{q\mu}{4\pi K_a h} \{\ln [x_1^2 + (x_2 + d)^2] + A \ln [x_1^2 + (x_2 - d)^2]\}$$

and also

$$\psi_b' = B \frac{q\mu}{4\pi K_a h} \ln [x_1^2 + (x_2 + d)^2] \tag{4-128}$$

Here B is another constant to be determined. Then at the discontinuity $x_2 = 0$, we have

$$1 + A = B \tag{4-129}$$

and

$$1 - A = \frac{K_b}{K_a} B \tag{4-130}$$

from the boundary conditions. Hence

$$B = \left(1 + \frac{K_b}{K_a}\right)^{-1} \tag{4-131}$$

and

$$A = \left(1 - \frac{K_b}{K_a}\right)\left(1 + \frac{K_b}{K_a}\right)^{-1} \tag{4-132}$$

Therefore, a point source of strength q at a distance $-d$ from the discontinuity is represented by

$$\psi_a' = \frac{q\mu}{4\pi K_a h} \left\{ \ln [x_1^2 + (x_2 + d)^2] + \frac{1 - (K_b/K_a)}{1 + (K_b/K_a)} \ln [x_1^2 + (x_2 - d)^2] \right\} \tag{4-133}$$

for $x_2 < 0$, and

$$\psi_b' = \frac{q\mu}{4\pi K_a h} \left(1 + \frac{K_b}{K_a}\right)^{-1} \ln (x_1^2 + (x_2 + d)^2] \tag{4-134}$$

for $x_2 > 0$.

Particularly extreme cases arise as one considers different values of K_b. For example, putting K_b equal to zero makes the surface of discontinuity an impermeable barrier. In this case ψ_b' is of no physical significance. On the other hand, letting K_b become infinite forces region (2) to be at zero potential everywhere. This makes the surface of discontinuity an equipotential surface.

Obviously, multiple wells can be treated by superposition, each well being represented in the same manner as the single well. Multiple regions of different permeability or multiple impermeable plane boundaries can be treated by extensions of this method.

Consider a single well located at a distance d from one boundary within an infinite strip of width L. The method of solution by images is most easily understood by considering an infinite array of such strips arranged side by side as shown in Figure 4-12.

An image well is placed in each strip as indicated. Note that the arrange-

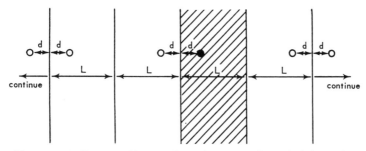

Figure 4–12. System of images for point source in an infinite strip.

ment of images is just that which would result by reflections in mirrors coincident with any adjacent pair of the boundaries, thus the terminology "image wells." Due to symmetry the impermeable boundaries can be removed and still no flow occurs across these lines. An infinite line of wells of equal strengths placed in a plane as shown would yield a potential distribution such that no flow across these lines would occur.

Further application of point sources and the method of images are treated in Chapter 7.

4.60: Gravity Drainage: Free Surfaces

In the field of hydrology the dominant mechanism of flow is that frequently referred to as gravity drainage. For example, flow of water through an earthen dam will occur as a result of the gravitational force on the fluid. If capillary effects are neglected, such flow gives rise to what is called a free surface or a surface which is simultaneously a streamline and a surface of constant pressure. This is illustrated as follows.

Consider an earthen reservoir as shown in Figure 4-13. The dam and reservoir are underlaid by an impermeable stratum. The water flows through the dam under the action of gravity.

Within the dam a surface separating that portion of the porous medium filled with air (and water vapor) and the portion filled with water must exist. Since the air is at uniform pressure it follows that the pressure has the same value everywhere on this surface. On the other hand, under

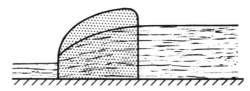

Figure 4–13. Seepage through an earthen dam showing a free surface and a surface of seepage.

steady-state conditions this surface is stationary; there is no component of fluid velocity normal to this surface. The fluid velocity is everywhere tangent to this surface and the surface is therefore a streamline. This is the free surface.

In general, the free surface intersects the downstream surface of the dam at some point which may be above the liquid surface of the downstream reservoir. The portion of the dam face between this point and the surface of the downstream reservoir represents a surface of seepage. This is a surface of constant pressure across which flow occurs.

In the absence of free surfaces and seepage surfaces the mathematical solution of gravity-flow problems is relatively straightforward. For example, in two dimensions, x_1 horizontal and x_3 vertical, the equation of flow is

$$\frac{\partial^2}{\partial x_1^2}(p + \rho g x_3) + \frac{\partial^2}{\partial x_2^2}(p + \rho g x_3) = 0 \qquad (4\text{-}135)$$

The porous medium is considered to be homogeneous and isotropic and the fluid of density ρ to be incompressible. For specified boundary conditions, this may be solved by the methods already discussed or other suitable methods.

If a free surface does exist, it is specified by the two conditions

$$\begin{cases} p = p_a \\ \dfrac{\partial}{\partial l_n}(p + \rho g x_3) = 0, \end{cases} \qquad \text{on free surface} \qquad (4\text{-}136)$$

Here p_a is the uniform pressure in the gas region and $\partial/\partial l_n$ denotes the space derivative normal to the free surface. The form of the free surface and the distribution of pressure must be determined simultaneously.

Methods for the solution of problems of this type are not highly developed. One method is based on the mapping of the region of flow in the hodograph plane, that is, the velocity plane. The application of this method to problems in hydrology is discussed by Muskat in his "Flow of Homogeneous Fluids Through Porous Media."[4]

Due to the complexity of problems of this type most practical studies of such problems have been carried out with models, either physical or analog. The theory of such models is described in Chapter 9.

EXERCISES

1. For steady horizontal flow in a plane stratum of uniform thickness employ the conformal mapping $w = z + z^{-1}$ to solve the problem of flow around a cylindrical obstacle of unit radius normal to the plane. Consider linear flow to exist at great distance from the impermeable obstacle.
2. Use the transformation $w = z^{-1}$ and the method of images to show that a plane

point sink located at x_0, y_0 in a plane region bounded by a circular impermeable barrier of radius R and center at (R, O) is represented by

$$\psi' = \frac{q\mu}{4\pi Kh} \ln \left\{ \frac{\left[\dfrac{x}{x^2 + y^2} - \dfrac{x_0}{x_0{}^2 + y_0{}^2} \right]^2 + \left[\dfrac{y}{x^2 + y^2} - \dfrac{y_0}{x_0{}^2 + y_0{}^2} \right]^2}{\left[\dfrac{x}{x^2 + y^2} - \dfrac{1}{R} + \dfrac{x_0}{x_0{}^2 + y_0{}^2} \right]^2 + \left[\dfrac{y}{x^2 + y^2} - \dfrac{y_0}{x_0{}^2 + y_0{}^2} \right]^2} \right\}$$

3. Show that for a single well in a uniform plane anisotropic stratum the curves of equal potential in the plane are confocal ellipses with ratio of major to minor axes given by K_1/K_2, where K_1 is permeability along major axis and K_2 is permeability along minor axis.

References

1. Churchill, R. V., "Introduction to Complex Variables and Applications," McGraw-Hill Book Co., New York (1948).
2. Gibbs, W. J., "Conformal Transformations in Electrical Engineering," Chapman and Hall, London (1958).
3. Jahnke, E., and Emde, F., "Tables of Functions with Formulas and Curves," Dover Pub., New York (1945).
4. Muskat, M., "Flow of Homogeneous Fluids through Porous Media," McGraw-Hill Book Co., New York (1937); J. W. Edwards, Inc., Ann Arbor (1946).

5. TRANSIENT LAMINAR FLOW OF HOMOGENEOUS FLUIDS

5.10: Transient Flow of Compressible Liquids

A vast number of important problems in the production of oil can be treated approximately in terms of the flow of homogeneous compressible liquids with negligible gravity effects.

From section 3.70 the differential equation describing this type of flow can be written as

$$\frac{\partial^2 \rho}{\partial x_1{}^2} + \frac{\partial^2 \rho}{\partial x_2{}^2} + \frac{\partial^2 \rho}{\partial x_3{}^2} = \frac{\phi \mu c}{K} \frac{\partial \rho}{\partial t} \tag{5-1}$$

for a fluid of constant compressibility flowing through an isotropic homogeneous porous medium.

If the medium is homogeneous but not isotropic, the differential equation has the form

$$K_1 \frac{\partial^2 \rho}{\partial x_1{}^2} + K_2 \frac{\partial^2 \rho}{\partial x_2{}^2} + K_3 \frac{\partial^2 \rho}{\partial x_3{}^2} = \phi \mu c \frac{\partial \rho}{\partial t} \tag{5-2}$$

or

$$\frac{\partial^2 \rho}{\partial \eta_1{}^2} + \frac{\partial^2 \rho}{\partial \eta_2{}^2} + \frac{\partial^2 \rho}{\partial \eta_3{}^2} = \frac{\phi \mu c}{K_1} \frac{\partial \rho}{\partial t} \tag{5-3}$$

in terms of the coordinates defined by

$$\eta_i = x_i \sqrt{\frac{K_1}{K_i}}, \qquad i = 1,2,3 \tag{5-4}$$

In equations (5-2) and (5-3) the coordinates are parallel to the principal axes of permeability. Note that, according to section 3-70, ρ can be replaced by p if the fluid is only slightly compressible.

A number of general techniques for the solution of such equations exists. Some of these are described and illustrated in the following sections. For a more complete discussion of methods of solution of these equations the reader may consult mathematical texts or the treatise on heat conduction by Carslaw and Jaeger.[4]

5.20 The Laplace Transform and Duhamel's Theorem

Duhamel's theorem, or Duhamel's formula, is so fundamental to the study of transient-flow phenomena that it is developed in some detail, and in various forms. The Laplace transform is also of great utility in transient-flow problems and offers a convenient means of deducing the several forms of Duhamel's formula. Thus, we present first the definition and basic properties of the Laplace transform.

The Laplace transform of a function $f(t)$ is defined as

$$L\{f(t)\} = \int_0^\infty e^{-st} f(t) \, dt = g(s) \tag{5-5}$$

provided the integral exists. The parameter s is usually called the transform variable.

The inverse transformation is defined in terms of a contour integral in the complex plane as

$$L^{-1}\{g(s)\} = \frac{1}{2\pi i} \lim_{\gamma \to \infty} \int_{c-i\gamma}^{c+i\gamma} e^{st} g(s) \, ds = f(t) \tag{5-6}$$

the particular contour being selected so that all poles of the function $g(s)$ are to the left of $s = c$.

In practice the complex inversion integral given by equation (5-6) rarely needs to be employed. Extensive tables of Laplace transforms[13] are available so that given a transform function, $g(s)$, one can usually find the corresponding function, $f(t)$, in a table.

The utility of the Laplace transform in problems of transient flow of homogeneous fluids arises from the following property.

If the integral in equation (5-5) is integrated by parts, there results

$$g(s) = \frac{f(0)}{s} + \frac{1}{s} \int_0^\infty e^{-st} \frac{df(t)}{dt} \, dt \tag{5-7}$$

The integral appearing here is the transform of df/dt. Rearranging gives:

$$L\left\{\frac{df(t)}{dt}\right\} = sg(s) - f(0) \tag{5-8}$$

Similar relations exist for higher derivatives.

Now consider a linear partial differential equation, in the variables x_1, x_2, x_3 and t, having a first partial derivative with respect to t of the dependent variable, ρ. Thus

$$D[\rho(x_1, x_2, x_3, t)] = \frac{\partial \rho}{\partial t} (x_1, x_2, x_3, t) \tag{5-9}$$

where D is a differential operator in the space variables, the left member of equation (5-2), for example.

Applying the Laplace transform operation to both sides of this equation yields

$$D[\bar{\rho}(x_1, x_2, x_3, s)] = s\bar{\rho}(x_1, x_2, a_3, s) - \rho(x_1, x_2, x_3, 0) \tag{5-10}$$

Here $\bar{\rho}(x_1, x_2, x_3, s)$ is the transform of the function $\rho(x_1, x_2, x_3, t)$ with respect to t. It is assumed here that the order of differentiation with respect to the space coordinates and integration with respect to t can be interchanged.

Whereas the original differential equation was a partial differential equation in the four variables x_1, x_2, x_3 and t we now have a differential equation in just three variables, x_1, x_2, and x_3. The quantity s is treated as a fixed parameter.

If the initial distribution of ρ is uniform, then

$$\rho(x_1, x_2, x_3, 0) = \rho_0 = \text{constant} \tag{5-11}$$

and

$$L\{\rho_0\} = \frac{\rho_0}{s} \tag{5-12}$$

In this case it is convenient to take as the dependent variable

$$Y = \rho - \rho_0 \tag{5-13}$$

Then equation (5-10) becomes

$$D[\bar{Y}(x_1, x_2, x_2, s)] = s\bar{Y}(x_1, x_2, x_3, s) \tag{5-14}$$

where $\bar{Y}$ is the Laplace transform of Y on the variable t.

Now suppose that the surface bounding the region of flow is composed of parts on which Y is zero, parts across which no flow occurs and one part on which Y is a specified function of t, say

$$Y = F(t) \text{ on surface } \sigma \tag{5-15}$$

Let us solve the flow problem first for $F(t) = 1$, and denote the transform of this solution by $\bar{Y}_1$.

Now $Y_1(x_1, x_2, x_3, t)$ satisfies all the boundary conditions of the original problem except on the surface σ. On σ

$$\bar{Y} = \bar{F}(s) \tag{5-16}$$

and

$$\bar{Y}_1 = \frac{1}{s} \tag{5-17}$$

Consider as a trial solution of the original problem the function whose transform is

$$\bar{Y}(x_1, x_2, x_3, s) = s\bar{F}(s)\bar{Y}_1(x_1, x_2, x_3, s) \tag{5-18}$$

This function satisfies the differential equation and the boundary condition on σ as well as all other boundaries. Hence this is the transform of the required solution.

It can be shown that the inverse transform of the right member is such that

$$Y(x_1, x_2, x_3, t) = \int_0^t F(t - \tau) \frac{dY_1(x_1, x_2, x_3, \tau)}{dt} d\tau \tag{5-19}$$

This is Duhamel's formula.[5] This shows that the solution of a problem with a variable boundary condition can be obtained from the solution of the same problem with Y being unity on this boundary. This is especially useful in certain problems of oil production.

In particular, if the fluid is a liquid of only slight compressibility, we have from Chapter 3

$$\frac{\partial^2 p}{\partial x_1^2} + \frac{\partial^2 p}{\partial x_2^2} + \frac{\partial^2 p}{\partial x_3^2} = \frac{\phi \mu c}{K} \frac{\partial p}{\partial t} \tag{5-20}$$

if the medium is homogeneous and isotropic. Introducing the new variable

$$P = p - p_0 \tag{5-21}$$

where p_0 is the initial uniform pressure in the region of flow, this becomes

$$\nabla^2 P = \frac{\phi \mu c}{K} \frac{\partial P}{\partial t} \tag{5-22}$$

Now let P_σ denote the value of P on the bounding surface σ. From equation (5-18), we have in this case

$$\bar{P}(x_1, a_2, x_3, s) = s\bar{P}_\sigma(s)\bar{P}_1(x_1, x_2, x_3, s) \tag{5-23}$$

Here $\bar{P}_1$ is the transform of the solution for $P_\sigma = 1$.

Differentiating both sides of this equation with respect to distance normal to σ, multiplying by $-K/\mu$ and integrating over σ yields

$$\bar{q}_\sigma(s) = s\bar{P}_\sigma(s) \left[-\frac{K}{\mu} \int \left(\frac{\partial \bar{P}_1}{\partial l_{n\sigma}} \right) d\sigma \right] \tag{5-24}$$

Here $q_\sigma(t)$ is the flow rate through σ for $P = P_\sigma(t)$ on σ.

An equation of exactly the same form can be written for $q_\sigma'(t)$ corresponding to $P = P_\sigma'(t)$ on σ. The factor in brackets would be the same. It follows that

$$\frac{s\bar{q}_\sigma(s)}{\bar{P}_\sigma(s)} = \frac{s\bar{q}_\sigma'(s)}{\bar{P}_\sigma'(s)} \tag{5-25}$$

or

$$s\bar{P}_\sigma'(s)\bar{q}_\sigma(s) = s\bar{P}_\sigma(s)q_\sigma'(s) \tag{5-26}$$

The inverse Laplace transform then yields

$$\int_0^t q_\sigma(\tau)\,\frac{dP_\sigma'(t-\tau)}{dt}\,d\tau = \int_0^t q_\sigma'(\tau)\,\frac{dP_\sigma(t-\tau)}{dt}\,d\tau \tag{5-27}$$

If P is a known function of t on σ, $P_\sigma(t)$, for a given flow rate $q_\sigma(t)$, one can use this equation to compute $P_\sigma'(t)$ which would exist on σ if the flow rate through σ were $q_\sigma'(t)$.

If q_σ' is a constant this equation takes the form

$$P_\sigma(t) = \frac{1}{q_\sigma'}\int_0^t q_\sigma(\tau)\,\frac{dP_\sigma'(t-\tau)}{dt}\,d\tau \tag{5-28}$$

This has applications in studies of pressure build-up tests in oil wells discussed in section 5.40.

Another useful result also follows from Duhamel's formula. Both sides of equation (5-26) may be divided by s to yield

$$s\bar{P}_\sigma'(s)\,\frac{\bar{q}_\sigma(s)}{s} = s\bar{P}_\sigma(s)\,\frac{\bar{q}_\sigma'(s)}{s} \tag{5-29}$$

The inverse transform then yields

$$\int_0^t P_\sigma'(t-\tau)\,\frac{dQ_\sigma(\tau)}{dt}\,d\tau = \int_0^t P_\sigma(t-\tau)\,\frac{dQ_\sigma'}{dt}(\tau)\,d\tau \tag{5-30}$$

Here

$$Q_\sigma(t) = \int_0^t q_\sigma(\tau)\,d\tau \tag{5-31}$$

and Q_σ' is similarly defined. That is, Q_σ' is the cumulative flow across σ when the pressure on σ is given by $P_\sigma'(t)$.

If $P_\sigma'(t)$ has the constant value, P_σ', then equation (5-30) can be written as

$$Q_\sigma(t) = \frac{1}{P_\sigma'}\int_0^t P_\sigma(t-\tau)\,\frac{dQ_\sigma'(\tau)}{dt}\,d\tau \tag{5-32}$$

This result has important applications in the study of natural water drive petroleum reservoirs. This is elaborated in section 5.50.

5.30: Sources and Sinks; the Method of Images

The representation of fluid sources and sinks by mathematical singularities, point sources, introduced in Chapter 4 is extended to problems of compressible fluids.

Consider a homogeneous isotropic porous medium of infinite extent which is filled with a compressible liquid. When time $t = \tau$ let a bit of fluid mass, δm, be added to the fluid at the point $x_i = x_i'$, $i = 1,2,3$.

The fluid density ρ must satisfy the differential equation (5-1). At time $t = \tau$, $\rho = \rho_0$ everywhere except at the point (x_1', x_2', x_3'). The function

$$\rho = \rho_0 + \frac{B}{(t - \tau)^{3/2}} \exp \left\{ \frac{-\phi\mu c[(x_1 - x_1')^2 + (x_2 - x_2')^2 + (x_3 - x_3')^2]}{4K(t - \tau)} \right\} \tag{5-33}$$

satisfies the differential equation. Also as $t \to \tau$ this function approaches ρ_0 at all points except (x_1', x_2', x_3').

The excess mass of fluid in the medium is

$$\delta m = \int_{-\infty}^{+\infty} \int_{-\infty}^{+\infty} \int_{-\infty}^{+\infty} \phi(\rho - \rho_0) \, dx_1 \, dx_2 \, dx_3 \tag{5-34}$$

When equation (5-33) is substituted for ρ in this equation there results

$$\delta m = 8\phi B \left(\frac{\pi K}{\phi\mu c} \right)^{3/2} \tag{5-35}$$

The solution

$$\rho - \rho_0 = \frac{\delta m}{\phi} \left[\frac{\phi\mu c}{4\pi K(t - \tau)} \right]^{3/2} \exp \left\{ \frac{-\phi\mu c[(x_1 - x_1')^2 + (x_2 - x_2')^2 + (x_3 - x_3')^2]}{4K(t - \tau)} \right\} \tag{5-36}$$

corresponds to a mass of fluid δm liberated in the medium at the point (x_1', x_2', x_3'), at time $t = \tau$. If fluid is being introduced continually at the point in question, then

$$\delta m = \dot{m}(\tau) \, d\tau \tag{5-37}$$

where $\dot{m}$ is the rate at which fluid mass is introduced. Thus, it follows that in this case

$$\rho - \rho_0 = \frac{1}{\phi} \left(\frac{\phi\mu c}{4\pi K} \right)^{3/2} \int_0^t \frac{\dot{m}(\tau)}{(t - \tau)^{3/2}} \exp \left(-\frac{\phi\mu c R^2}{4K(t - \tau)} \right) d\tau \tag{5-38}$$

Here the notation

$$R^2 = (x_1 - x_1')^2 + (x_2 - x_2')^2 + (x_3 - x_3')^2 \tag{5-39}$$

has been used.

For the particular case in which $\dot{m}$ is a constant, this takes the form*

$$\rho - \rho_0 = \frac{\dot{m}\mu c}{4\pi KR} \text{ erfc} \sqrt{\frac{\phi\mu cR^2}{4Kt}} \tag{5-40}$$

which represents a point source of strength $\dot{m}$ at the point (x_1', x_2', x_3').

The point source in the plane is deduced from this result as an infinite uniform line source in three dimensions. Consider fluid introduced into the medium at the constant mass rate $\dot{m}'$ per unit length along the line $x_1 = x_1'$, $x_2 = x_2'$. By addition of solutions of the form of equation (5-40), there results

$$\rho - \rho_0 = \int_{-\infty}^{+\infty} \frac{\dot{m}'\mu c}{4\pi KR} \text{ erfc} \sqrt{\frac{\phi\mu cR^2}{4Kt}} \, dx_3' \tag{5-41}$$

which can be shown to be

$$\rho - \rho_0 = \frac{\dot{m}'\mu c}{4\pi K} \left[-Ei \left(-\frac{\phi\mu cr^2}{4Kt} \right) \right] \tag{5-42}$$

Here

$$r^2 = (x_1 - x_1')^2 + (x_2 - x_2')^2 \tag{5-43}$$

and

$$-E_i(-x) = \int_x^\infty \frac{e^{-\lambda}}{\lambda} \, d\lambda \tag{5-44}$$

is the exponential integral function.

For small values of the argument this function can be approximated as

$$-Ei(-x) \approx -\gamma - \ln x \tag{5-45}$$

where $\gamma = 0.5772$ is Euler's constant.

When the liquid is only slightly compressible the point sources can be expressed in terms of the pressure p. Since

$$\rho = \rho_0 e^{c(p-p_0)} \tag{5-46}$$

it follows that for small c and small values of $p - p_0$

$$\rho - \rho_0 \approx c\rho_0(p - p_0) \tag{5-47}$$

Also the mass flow rate can be expressed in terms of the volume rate of

* Here erfc denotes the complementary error function defined by:

$$\text{erfc } x = \frac{2}{\sqrt{\pi}} \int_x^\infty e^{-\lambda^2} \, d\lambda$$

flow. For example, let the volume rate of flow per unit length be q/h for the line source, as measured at the original pressure p_0. Then

$$\dot{m}' = \frac{\rho_0 q}{h} \tag{5-48}$$

and the line source is represented by

$$p - p_0 = \frac{q\mu}{4\pi K h}\left[-Ei\left(-\frac{\phi\mu c r^2}{4Kt}\right)\right] \tag{5-49}$$

Note that this describes a line source normal to a stratum of uniform thickness h bounded above and below by impermeable boundaries, the flow rate from the line being q. If q is replaced by $-q$ this is an approximate representation of an oil well producing at constant rate q from a uniform stratum by fluid expansion.

The point sources described above were restricted to isotropic media. Similar treatment is possible for homogeneous anisotropic media. For example, the point source in the plane is in the anisotropic case

$$p - p_0 = \frac{q\mu}{4\pi h\sqrt{K_1 K_2}}\left\{-Ei\left[-\frac{\phi\mu c}{4t}\left(\frac{(x_1 - x_1')^2}{K_1} + \frac{(x_2 - x_2')^2}{K_2}\right)\right]\right\} \tag{5-50}$$

Here it is assumed that the liquid is only slightly compressible and the coordinate axes coincide with the principal axes of permeability, the permeabilities being K_1 and K_2, respectively.

As in the case of steady flow, these point sources represent mathematical singularities at the source point. Thus, for example, equation (5-49) yields $p - p_0 \to \infty$ as the source point is approached. Even so, such point sources have great utility in practical problems. Equation (5-49), for example, with q negative is a very good representation of a circular well of some radius r_w provided $r > r_w$ and

$$\frac{\phi\mu c r^2}{4Kt} < 0.25$$

This is established by comparison with the exact analytical solution for a well of finite radius.[11]

The fact that point sources represent mathematical singularities makes possible the superposition of such point sources to obtain solutions of the flow equations corresponding to several diffrent sources. That is, the presence of one point source in the region of flow has no effect on the flow rate of another point source.

Thus the distribution of pressure in a uniform isotropic medium containing a slightly compressible liquid having a line source of strength q_a/h per unit length beginning at time $t = \tau_a$, and another of strength q_b/h per unit length parallel to the first beginning at time $t = \tau_b$ is, for $\tau_b > \tau_a$

$$
p - p_0 = \begin{cases} \dfrac{q_a\mu}{4\pi Kh}\left\{-Ei\left[-\dfrac{\phi\mu c r_a^2}{4K(t-\tau_a)}\right]\right\}, & \tau_a < t < \tau_b \\[2em] \dfrac{q_a\mu}{4\pi Kh}\left\{-Ei\left[-\dfrac{\phi\mu c r_a^2}{4K(t-\tau_a)}\right]\right\} + \dfrac{q_b\mu}{4\pi Kh}\left\{-Ei\left[-\dfrac{\phi\mu c r_b^2}{4K(t-\tau_b)}\right]\right\}, & \tau_b < t \end{cases}
$$

(5-51)

Here r_a and r_b are the distances from line source a to the point at which p exists and from line source b to this point, respectively.

Taking $\tau_a = \tau_b$ and $q_a = q_b$, we see that the distribution of pressure is symmetric about the plane which bisects the normal line connecting the two line sources. Furthermore, the space derivative of p normal to this plane is zero. This plane is equivalent to an impermeable barrier. The source b is then the image of source a across this plane. This procedure can be extended to construct multiple images corresponding to multiple boundaries in much the same manner as was done for the steady-flow case.[14]

The superposition of sources employed above can also be employed as purely superposition in time to represent a well having step-wise variations in rate. Thus, in equation (5-51) let the two sources coincide in space, that is $r_a = r_b$. Then we have a representation of a single source with

$$q = 0,\ 0 < t < \tau_a$$

$$q = q_a,\ \tau_a < t < \tau_b$$

$$q = q_a + q_b,\ \tau_b < t$$

By introducing any number of sources at the point with suitable strengths, either positive or negative, we can represent any step-wise rate history desired. This result also follows from Duhamel's formula and has applications in the study of pressure build-up tests in oil wells.

5.40: Pressure Build-up Tests in Oil Wells

Pressure build-up tests are conducted in oil wells for several purposes: to determine the static pressure in the region of the well, to estimate the permeability of the formation being drained by the well and in conjunction with drawdown tests to estimate the extent of any permeability reduction in the vicinity of the well. Attempts have also been made to estimate from such tests other factors, such as the proximity of fault boundaries, or the porosity of the formation.

In this section the elementary theory describing such tests is developed following the work of Horner[3] and a few examples of a more complex nature are included to illustrate the fact that many factors have a bearing on the

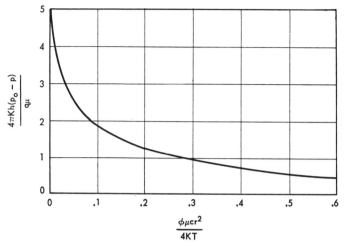

Figure 5-1. Pressure distribution about a circular well in an infinite plane stratum.

results of a pressure build-up test, far too many to permit an unambiguous interpretation of such tests in most cases.

Single Well in Infinite System. First we consider a homogeneous isotropic medium of uniform thickness h, porosity ϕ and permeability K containing a liquid of small compressibility c and viscosity μ.

At time $t = 0$ a well which completely penetrates the formation starts to produce fluid from the formation at a constant rate q. The upper and lower boundaries are assumed impermeable and for the moment the formation will be considered of infinite areal extent.

We represent the producing well by a line sink of strength $-q/h$ per unit length, thus

$$p(r, t) = p_0 - \frac{q\mu}{4\pi Kh}\left[-Ei\left(-\frac{\phi\mu cr^2}{4Kt}\right)\right] \tag{5-52}$$

where p_0 is the initial pressure in the formation and r is the radial distance from the axis of the well. This distribution of pressure has the form indicated in the graph in Figure 5-1.

At the well bore, $r = r_w$, this can be written in the approximate form

$$p_w = p_0 - \frac{q\mu}{4\pi Kh}\left[\ln\frac{4Kt}{\phi\mu cr_w^2} - \gamma\right] \tag{5-53}$$

for

$$\frac{4Kt}{\phi\mu cr_w^2} > 2 \tag{5-54}$$

by using the logarithmic approximation for the exponential integral function. This shows that the flowing pressure in the well declines as the logarithm of time.

If at some time t_s the well is shut in, the fluid in the formation will redistribute itself so as to establish a uniform pressure in the formation. The pressure as a function of r and t during the shut-in period is represented by the superposition of a line source of strength q/h per unit length on the already existing sink starting at time $t = t_s$. Thus, at the well bore

$$p_w = p_0 - \frac{q\mu}{4\pi Kh}\left[-Ei\left(-\frac{\phi\mu c r_w{}^2}{4Kt} \right) + Ei\left(-\frac{\phi\mu c r_w{}^2}{4K(t - t_s)} \right) \right] \qquad (5\text{-}55)$$

during the shut-in period, $t > t_s$.

Applying the logarithmic approximation to both exponential integrals now yields

$$p_w = p_0 - \frac{q\mu}{4\pi Kh} \ln \frac{t}{t - t_s} \qquad (5\text{-}56)$$

Or defining the shut-in time as

$$\delta t = t - t_s \qquad (5\text{-}57)$$

$$p_w = p_0 - \frac{q\mu}{4\pi Kh} \ln \frac{t_s + \delta t}{\delta t} \qquad (5\text{-}58)$$

which is valid for sufficiently large δt.

This shows that a plot of p_w versus the natural log of $\dfrac{(t_s + \delta t)}{\delta t}$ yields a straight line of slope $-q\mu/4\pi Kh$. If q, μ, and h were known quantities one could, from measurements of p_w as a function of t during such a shut-in test, compute the value of K from the slope of this line. However, in the practical application of this result many precautions must be observed. Note that the porous stratum must be uniform and of infinite extent, the fluid must be a homogeneous slightly compressible liquid and the well must have been produced at a constant rate prior to shut-in. Also the well is shut-in at the formation, not at the surface which may be several thousand feet above the formation. Obviously, all of these requirements cannot be met, particularly the requirement that the stratum be infinite in extent. However, for small t_s, boundaries may produce only negligible effects.

Well Near a Plane Barrier. To illustrate the manner in which various factors may produce deviations from this simple behavior consider the cases in which the uniform porous stratum terminates at a plane impermeable barrier near the well, say at a distance l from the well.

Applying the method of images an image well having exactly the same production history as the real well must be placed at a distance l on the side of the barrier away from the real well on the normal to the barrier passing through the well.

This gives for the well pressure during the shut-in period

$$p_w = p_0 + \frac{q\mu}{4\pi Kh} \ln \frac{\delta t}{t_s + \delta t} + \frac{q\mu}{4\pi Kh} \left[-Ei\left(-\frac{\phi\mu c l^2}{Kt}\right) + Ei\left(-\frac{\phi\mu c l^2}{Kt_s}\right) \right] \quad (5\text{-}59)$$

for moderate values of δt. Here l^2 is so large that for moderate values of δt the logarithmic approximation is not applicable. The exponential integrals appearing here change relatively slowly with δt for moderate δt. These terms are nearly constant and for moderate values of δt p_w plotted versus

$$\ln \frac{t_s + \delta t}{\delta t}$$

again yields a straight line of slope $-q\mu/4\pi Kh$.

However, for large values of δt the logarithmic approximation can be applied to these terms to yield

$$p_w = p_0 + 2\,\frac{q\mu}{4\pi Kh} \ln \frac{t_s + \delta t}{\delta t} \quad (5\text{-}60)$$

for large δt. On semi-log paper this is a straight line having twice the slope of the line corresponding to small values of δt. (Figure 5-2.)

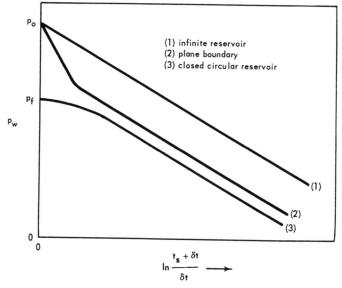

Typical Pressure Build-up Curves

Figure 5–2. Typical pressure build-up curves.

This change in slope is characteristic for a system of this geometry and can be used to detect the presence of such a plane "fault" boundary. Furthermore, detailed analysis shows that this change in slope occurs at a value of δt which is dependent on l, the distance to the boundary. This then offers a method of estimating l from such build-up data.

Bounded System. In both of the cases considered above the ultimate value of p_w for $\delta t \to \infty$ is just p_0, the original pressure in the stratum. This is because the system is of infinite extent in both cases. Obviously, the behavior of a finite system must be quite different. A single well in a uniform stratum of thickness h bounded by a closed impermeable boundary producing at a constant rate q will in time t_s produce a volume of fluid

$$Q = qt_s \tag{5-61}$$

This volume is supplied by expansion of the fluid within the reservoir. Thus

$$qt_s = c(p_0 - p_f)V_p \tag{5-62}$$

where p_f is the final uniform shut-in pressure in the system and V_p is the pore volume of the reservoir.

Thus, even though the pressure build-up curve may follow the simple theory outlined up to relatively large values of δt it must break away from the straight line described to approach the value

$$p_f = p_0 - \frac{qt_s}{cV_p} \tag{5-63}$$

as a limiting value.

The analytical solutions of the flow equation for wells in bounded systems have been worked out for several cases. Hurst and Van Evergingen[11] have published the solution for the case of a well of radius r_w at the center of a circular reservoir of radius, r_e. The reservoir is assumed of uniform thickness h and porosity ϕ; the fluid has compressibility c and viscosity μ, and the well produces at the constant volumetric rate q, as measured at the original pressure p_0. This solution yields for the pressure at the well, p_w

$$p_w = p_0 - \frac{q\mu}{\pi Kh(R^2 - 1)}\left(\frac{1}{4} + \frac{Kt}{\phi\mu cr_w^2}\right)$$

$$- \frac{q\mu}{2\pi Kh}\left\{\frac{3R^4 - 4R^4 \ln R - R^2 - 1}{4(R^2 - 1)^2} + 2\sum_{n=1}^{\infty}\frac{J_1^2(\beta_n R)\ \exp\left(-\beta_n^2\ \dfrac{Kt}{\phi\mu cr_w^2}\right)}{\beta_n^2[J_1^2(\beta_n R) - J_1^2(\beta_n)]}\right. \tag{5-64}$$

Here R denotes r_e/r_w, J_1 is the Bessel function of the first kind of order one and the β_n are roots of the equation

$$J_1(\beta_n R)Y_1(\beta_n) - J_1(\beta_n)Y_1(\beta_n R) = 0 \tag{5-65}$$

Y_1 denoting the Bessel function of the second kind of order one.

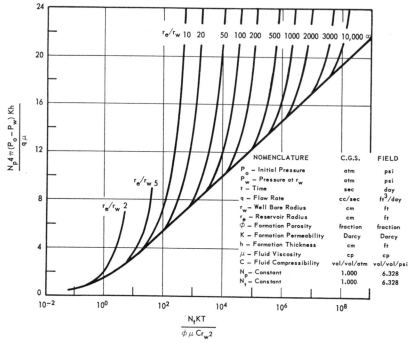

Figure 5–3. Compressible liquid flow; flowing pressure in a well at the center of a circular reservoir. (*After Hurst and Van Everdingen, 1949.*)

This solution shows that for very large values of t the flowing pressure in the well (at $r_w = r$) declines linearly with time.

Figure 5-3 shows the behavior of p_w with time for various values of r_e/r_w. These curves show quite clearly the effect of a boundary on flowing pressure. The straight-line portion corresponds to an infinite system. Thus, in all cases, the well behaves as though the reservoir were infinite up to some t which is determined by the value of r_e/r_w.

Duhamel's formula can be employed to construct the pressure build-up curve obtained after the well is shut-in. Thus, in equation (5-28), suppose that

$$q_\sigma = \begin{cases} q_\sigma', & 0 < t < t_s \\ 0, & t_s < t \end{cases} \tag{5-66}$$

then

$$P_\sigma(t) = \int_0^{t_s} \frac{dP_\sigma'\,(t-\tau)}{dt}\,d\tau \tag{5-67}$$

or

$$P_\sigma(t) = P_\sigma'(t) - P_\sigma'(t - t_s) \tag{5-68}$$

In terms of our p_w', this is

$$p_w = p_0 + p_w'(t) - p_w'(t - t_s) \tag{5-69}$$

where $p_w'(t)$ corresponds to continuing constant rate. Thus, a simple superposition of solutions yields the desired build-up curve. Such a curve is shown in Figure 5-2.

This build-up curve shows quite clearly the effect of the closed boundary. Indeed, the asymptotic value is just that predicted by equation (5-63).

Variable Production Rates. Pressure build-up curves for a well having variations in production rate prior to shut-in can be constructed from the solution corresponding to a constant continuing production rate by employing Duhamel's formula.

Thus, suppose that $p_w'(t)$ corresponds to the constant production rate $-q_w'$. Then for any production rate $-q_w(t)$ we have from equation (5-28)

$$p_w = \int_0^t \frac{q_w(\tau)}{q_w'} \frac{dp_w'(t-\tau)}{dt} \, d\tau \tag{5-70}$$

for the well pressure corresponding to the variable rate.

If the well is shut in at time t_s then $q_w(t)$ is zero for $t > t_s$, and this takes the form

$$p_w(t) = \int_0^{t_s} \frac{q_w(\tau)}{q_w'} \frac{dp_w'(t-\tau)}{dt} \, d\tau, \qquad t > t_s \tag{5-71}$$

for the pressure history during shut-in. This formulation applies without regard to the geometry of the well or reservoir.

The behavior of the flowing pressure when the well is reopened following a shut-in period can be obtained by an obvious extension of the above procedure, i.e., $q_w \neq 0$ for $t > t_0 > t_s$.

Limitations of Simple Theory. The few examples of pressure build-up in oil wells discussed in the preceding sections should not be applied without due consideration of their limitations. Many important factors have been neglected in order to illustrate the main characteristics. A more adequate treatment is possible but it would form a book in itself. Effects due to non-uniform thickness, non-uniform permeability, shutting in at the surface instead of at the sandface and multiphase flow can be handled, but this is beyond the scope of this volume. Some of these problems have been treated in the literature.[15, 16, 17]

5.50: Performance of Water-Drive Reservoirs

Many oil reservoirs produce oil (and gas) by a mechanism termed water drive. Often this is called natural water drive to distinguish it from artificial water drive which involves the injection of water into the oil-bearing formation as is discussed in Chapter 7.

Natural water drive arises because most oil reservoirs are in direct contact with large aquifers, or water-filled porous strata. As the pressure in the oil reservoir is lowered by the production of oil, expansion of the remaining oil occurs and also expansion of the water in the aquifer. There is an influx or flow of water into the region of the porous stratum originally filled primarily with oil. This is a simplified description of the natural water drive process.

An important problem for the reservoir engineer is the estimation of the extent of the aquifer and the prediction of the future decline of average reservoir pressure for a given schedule of oil production. This is so because the reduction in pressure gives rise to the evolution of gas from the oil with resultant reduction in the oil-production capacities of the wells in the reservoir.

Even though water is only very slightly compressible the very great extent (billions of barrels) of aquifers gives rise to sizable influxes. Hence water influx is a dominant factor in the control of reservoir pressure.

A variety of methods has been developed for treating the problem of water influx. All of these have the same physical and mathematical basis, namely that the aquifer can be described in terms of flow of a slightly compressible liquid through a porous medium. To illustrate the general method of treatment of this problem, we employ the following conceptual model.

Consider a reservoir containing oil, water and gas (some gas in solution in the oil). The free gas may be partly or completely segregated from the liquids, the segregated portion occupying a volume termed the gas cap. This reservoir has contact at a surface σ with a very large region of porous medium filled with water. The geometry and properties of this water region (the aquifer) are unknown.

The pressure in the aquifer can be described by a differential equation of the form

$$D[p(x_1, x_2, x_3, t)] = \frac{\partial p}{\partial t} \tag{5-72}$$

Here D denotes some differential operator. The form of this operator depends on the properties of the aquifer. In any event if gravity effects are negligible, this equation is linear and homogeneous in p. Thus Duhamel's formula can be applied.

It is assumed that the pressure $p_r(t)$ is uniform throughout the oil reservoir so that the pressure on the surface σ is identical to $p_r(t)$ at all times.

Now the equation describing the aquifer can in principle be solved for either constant pressure on σ or constant flow rate across σ. Suppose it is solved for constant pressure on σ with an initial uniform pressure p_0, everywhere. If it is solved for

$$p_0 - p_\sigma = 1 \tag{5-73}$$

then from this solution we can compute the cumulative influx of water across σ as $Q_1(t)$. Then, by Duhamel's formula in the form of equation (5-32)

$$Q_\sigma(t) = \int_0^t [p_0 - p_r(t - \tau)] \frac{dQ_1(\tau)}{dt} \, d\tau \tag{5-74}$$

is the cumulative influx which results for $p_\sigma = p_r(t)$.

If the size of the reservoir is known, as well as its initial contents, and the physical properties of its fluids are known then a volumetric balance can be written

$$Q_o + Q_w + Q_g = f(p_r) + Q_\sigma(t) \tag{5-75}$$

Here Q_o, Q_w and Q_g are the cumulative withdrawals of oil, water and gas, respectively, from the reservoir. $f(p_r)$ is a function of p_r which represents the net expansion of reservoir fluids due to the reduction in reservoir pressure.

For a period of several years data may be obtained on Q_o, Q_w, Q_g and p_r. Thus, from $t = 0$ to the current time, t_c, these are essentially known functions of time. Combining these two equations

$$Q_o + Q_w + Q_g - f(p_r) = \int_0^t [p_0 - p_r(t - \tau)] \frac{dQ_1(\tau)}{dt} \, d\tau \tag{5-76}$$

for $0 < t < t_c$, is obtained.

The procedure to be followed is to construct a function $Q_1(t)$ which satisfies this equation for $0 < t < t_c$. When such a function has been found it may be extrapolated for $t > t_c$ and equation (5-76) can then be used for predictions of $p_r(t)$ for specified withdrawals from the reservoir.

Note that the function $Q_1(t)$ is determined solely by the properties of the aquifer.

The other procedure for treating the aquifer problem is to consider the solution of the differential equation of the aquifer for constant rate of influx across σ. In particular, if the solution for

$$q_\sigma = 1$$

is obtained, then we denote the drawdown at σ by

$$[p_0 - p_\sigma(t)]_{q_\sigma=1} = F(t) \tag{5-77}$$

This is called the *influence function* of the aquifer.

Duhamel's formula in the form of equation (5-28) yields

$$p_0 - p_r(t) = \int_0^t q_\sigma(\tau) \frac{dF(t - \tau)}{dt}\, d\tau \tag{5-78}$$

Then, by differentiating the volumetric balance equation (5-75) with respect to time, there results

$$q_0 + q_w + q_g = \frac{df(p_r)}{dt} + q_\sigma(t) \tag{5-79}$$

Here q_o, q_w and q_g are the withdrawal rates of oil, water and gas, respectively.

Again data on q_o, q_w, q_g and p_r are available for $0 < t < t_c$ and hence $q_\sigma(t)$ is known for this time interval. The problem is to construct a function $F(t)$ satisfying equation (5-78) for $0 < t < t_c$. This function can then be extrapolated for $t > t_c$ and equations (5-78) and (5-79) used for prediction of future reservoir behavior.

The methods used to estimate the functions Q_1 and F are many and varied.[7, 10] To gain some insight into the nature of these functions several simple cases are described in following sections.

5.51: The Infinite Linear Aquifer

The simplest case of compressible liquid flow which can be applied to the aquifer problem is that of linear flow in a semi-infinite homogeneous medium. The mathematical description of this flow regime takes the form

$$\frac{\partial^2 p}{\partial x^2} = \frac{\phi \mu c}{K} \frac{\partial p}{\partial t} \tag{5-80}$$

with

$$\left.\begin{array}{c} p(x, 0) = p_0 \\[1em] -\dfrac{KA}{\mu} \dfrac{\partial p(0, t)}{\partial x} = q_\sigma = \text{constant} \\[1em] \lim_{x \to \infty} p(x, t) = p_0 \end{array}\right\} \tag{5-81}$$

This is reduced to a much simpler form by introducing the dimensionless variables,

$$\zeta = \frac{x}{\sqrt{A}}, \quad \lambda = \frac{Kt}{\phi \mu c A}, \quad \text{and} \quad \mathcal{P} = \frac{(p_0 - p)K\sqrt{A}}{q_\sigma \mu} \tag{5-82}$$

Thus

$$\frac{\partial^2 \mathcal{P}}{\partial \zeta^2} = \frac{\partial \mathcal{P}}{\partial \lambda} \tag{5-83}$$

With the boundary and initial conditions

$$\left. \begin{array}{l} \mathcal{P}(\zeta, 0) = 0 \\[2mm] \dfrac{\partial \mathcal{P}}{\partial \zeta}(0, \lambda) = 1 \\[2mm] \lim_{\zeta \to \infty} \overline{\mathcal{P}}(\zeta, \lambda) = 0 \end{array} \right\} \tag{5-84}$$

Taking the Laplace transform of equation (5-83) with respect to λ then yields

$$\frac{d^2 \overline{\mathcal{P}}}{d\zeta^2}(\zeta, s) = s\overline{\mathcal{P}}(\xi, s) \tag{5-85}$$

The total derivative can be used since $\overline{\mathcal{P}}(\xi, s)$ is a function of ζ only. The characteristic solutions of this equation are

$$e^{\zeta \sqrt{s}} \quad \text{and} \quad e^{-\zeta \sqrt{s}}$$

In view of the last of the boundary conditions in (5-84) the negative exponent must be used. Thus, take

$$\mathcal{P}(\zeta, s) = Be^{-\zeta \sqrt{s}} \tag{5-86}$$

where B is a constant.

Taking the Laplace transform of the boundary condition at $\zeta = 0$ yields

$$\frac{\partial \overline{\mathcal{P}}}{\partial \zeta}(0, s) = s^{-1} \tag{5-87}$$

Then, substituting the solution above, yields

$$B = -s^{-3/2} \tag{5-88}$$

Thus

$$\overline{\mathcal{P}}(\zeta, s) = -\frac{1}{s^{3/2}} e^{-\zeta \sqrt{s}} \tag{5-89}$$

is the Laplace transform of the desired solution.

From a table of Laplace transforms, we obtain

$$L^{-1}\left\{ \frac{1}{\sqrt{s}} e^{-\zeta \sqrt{s}} \right\} = \frac{1}{\sqrt{\pi \lambda}} e^{-(\zeta^2/4\lambda)} \tag{5-90}$$

and also for any function $H(\lambda)$

$$L^{-1}\left\{\frac{1}{s}h(s)\right\} = \int_0^\lambda H(\eta)\,d\eta \tag{5-91}$$

where

$$L^{-1}\{h(s)\} = H(\lambda) \tag{5-92}$$

Thus

$$\mathcal{P}(\zeta, \lambda) = -\int_0^\lambda \frac{1}{\sqrt{\pi\tau}}\,e^{-(\zeta^2/4\tau)}\,d\tau \tag{5-93}$$

is the desired solution of the flow problem.

This solution can be put into a more convenient form by a change of variables in the above integral. Let

$$\eta^2 = \frac{\zeta^2}{4\tau}, \quad \text{or} \quad \eta = \frac{\zeta}{2\sqrt{\tau}} \tag{5-94}$$

then we obtain

$$\mathcal{P}(\zeta, \lambda) = \frac{\zeta}{\sqrt{\pi}}\int_{\zeta/2\sqrt{\lambda}}^\infty e^{-\eta^2}\frac{d\eta}{\eta^2} \tag{5-95}$$

Integration by parts then gives*

$$\mathcal{P}(\zeta, \lambda) = 2\sqrt{\frac{\lambda}{\pi}}\,e^{-(\zeta^2/4\lambda)} - \frac{2\zeta}{\sqrt{\pi}}\int_{\zeta/2\sqrt{\lambda}}^\infty e^{-\eta^2}d\eta \tag{5-96}$$

Thus, at $\zeta = 0$, that is on the surface σ, we have

$$\mathcal{P}(0, \lambda) = 2\sqrt{\frac{\lambda}{\pi}} \tag{5-97}$$

or

$$p_c - p_\sigma = \frac{2q_\sigma\mu}{K\sqrt{A}}\sqrt{\frac{Kt}{\pi\phi\mu cA}} \tag{5-98}$$

or

$$p_\sigma = p_0 - \frac{2q_\sigma}{A}\sqrt{\frac{\mu t}{\pi\phi cK}} \tag{5-99}$$

* The integral appearing here is the complementary error function:

$$\text{erfc } \gamma = \frac{2}{\sqrt{\pi}}\int_\gamma^\infty e^{-\eta^2}\,d\eta = 1 - \text{erf } \gamma$$

The form, (5-98), with $q_\sigma = 1$ represents the influence function for this aquifer. Thus, if $q_\sigma = q_\sigma(t)$ is variable we have by Duhamel's formula (Equation 5-78)

$$p_\sigma = p_0 - \frac{1}{A} \sqrt{\frac{\mu}{\pi\phi cK}} \int_0^t q_\sigma(\tau) \frac{d\tau}{\sqrt{t - \tau}} \qquad (5\text{-}100)$$

The case of constant p_σ can be derived from this.

Taking the Laplace transform of both sides with respect to t yields for $p_\sigma = $ constant

$$\frac{p_0 - p_\sigma}{s} = \frac{2}{A} \sqrt{\frac{\mu}{\pi\phi cK}} \, \bar{q}_\sigma \frac{\sqrt{\pi}}{\sqrt{s}} \qquad (5\text{-}101)$$

Thus

$$\bar{q}_\sigma = \frac{A}{2} \sqrt{\frac{\phi cK}{\pi\mu}} (p_0 - p_\sigma) \frac{\sqrt{\pi}}{\sqrt{s}} \qquad (5\text{-}102)$$

and the inverse transform then yields

$$q_\sigma(t) = \frac{A}{2} \sqrt{\frac{\phi cK}{\pi\mu}} (p_0 - p_\sigma) \frac{1}{\sqrt{t}} \qquad (5\text{-}103)$$

This shows that initially, at $t = 0$, the flow rate from the aquifer would be infinite if at $t = 0$, a constant drawdown was imposed on the outflow surface σ.

Integrating with respect to t gives

$$Q_\sigma(t) = A \sqrt{\frac{\phi cK}{\pi\mu}} (p_0 - p_\sigma) \sqrt{t} \qquad (5\text{-}104)$$

for the cumulative influx from the aquifer for constant $p_0 - p_\sigma$. Putting $p_0 - p_\sigma = 1$ yields what we called $Q_1(t)$ in the previous section for this type of aquifer. For variable p_σ we then have from equation (5-74)

$$Q_\sigma(t) = \frac{A}{2} \sqrt{\frac{\phi cK}{\pi\mu}} \int_0^t [p_0 - p_0(t - \tau)] \frac{d\tau}{\sqrt{\tau}} \qquad (5\text{-}105)$$

for the cumulative influx from the infinite linear aquifer.

5.52: The Infinite Radial Aquifer

The solution of the equation of flow for a liquid of slight compressibility in a plane radial system of thickness h, inner radius r_w and outer radius r_e was given in equation (5-64), for $r = r_w$. This solution represents the pressure on the inner surface $\sigma(r = r_w)$ for constant influx rate q from a uniform aquifer.

If only a portion of the full circle is considered, a radial segment of angular width α measured in radians, then the factor π must everywhere be replaced by $\alpha/2$ in this equation.

This representation of the radial aquifer is not very useful because of its complexity. For the case of $r_e \to \infty$ an approximate representation of $p_w(t)$ can be constructed.

For very small t, $p_w(t)$ should be approximately the same as for a linear aquifer, i.e. equation (5-99) should apply for small t. For very large values of t, $p_w(t)$ should be approximated by the point sink solution, (the straight line in Figure 5-3), or equation (5-53). A function having these characteristics and in good numerical agreement with the exact analytical solution shown in Figure 5-3 for $r_e/r_w \to \infty$ is: ($\gamma = 0.5772$)

$$p_w \approx p_0 + \frac{q\mu}{2\alpha Kh} \left\{ 2\left[1 - \exp\left(-\frac{2}{\sqrt{\pi}} \sqrt{\frac{Kt}{\phi\mu cr_w{}^2}} \right) \right] + \ln\left(1 + \frac{4Kt}{\gamma e^2 \phi\mu cr_w{}^2} \right) \right\} \quad (5\text{-}106)$$

This representation with $q = 1$ can be used to write the influence function for an infinite uniform radial aquifer as

$$F(t) \approx \frac{\mu}{2\alpha Kh} \left\{ 2\left[1 - \exp\left(-\frac{2}{\sqrt{\pi}} \sqrt{\frac{Kt}{\phi\mu cr_w{}^2}} \right) \right] + \ln\left(1 + \frac{Kt}{\gamma e^2 \phi\mu cr_w{}^2} \right) \right\} \quad (5\text{-}107)$$

From this one can obtain $Q_1(t)$ as was done for the linear aquifer but the analysis is more complex.

5.53: The Tilted Aquifer

The preceding discussion of aquifers applies only for horizontal aquifers, or the neglect of all effects of gravity. The exact formulation for a two-dimensional system (x_1 horizontal and x_3 vertical) yields by Darcy's law, the equation of state and the equation of continuity

$$\frac{\partial}{\partial x_1}\left[\rho^2 \frac{\partial\psi}{\partial x_1} \right] + \frac{\partial}{\partial x_3}\left[\rho^2 \frac{\partial\psi}{\partial x_3} \right] = \frac{\phi\mu}{K} \frac{\partial\rho}{\partial t} \quad (5\text{-}108)$$

Here the medium is assumed homogeneous and isotropic and ψ is given by

$$\psi = \int_{p_0}^{p} \frac{dp}{\rho(p)} + gx_3 \quad (5\text{-}109)$$

If the aquifer is treated as a thin stratum at an angle α with respect to the horizontal, then we neglect variations in p and ρ across its thickness and introduce the length variable x measured parallel to the stratum. Thus

$$x_1 = x \cos \alpha - y \sin \alpha$$
$$x_3 = x \sin \alpha + y \cos \alpha \quad (5\text{-}110)$$

where y is measured perpendicular to the stratum. In terms of x and y, we have

$$\frac{\partial}{\partial x}\left[\rho^2 \frac{\partial \psi}{\partial x}\right] + \frac{\partial}{\partial y}\left[\rho^2 \frac{\partial \psi}{\partial y}\right] = \frac{\phi \mu}{K} \frac{\partial \rho}{\partial t} \tag{5-111}$$

and

$$\psi = \int_{p_0}^{p} \frac{dp}{\rho(p)} + g(x \sin \alpha + y \cos \alpha) \tag{5-112}$$

Then

$$\frac{\partial \psi}{\partial x} = \frac{1}{\rho} \frac{\partial p}{\partial x} + g \sin \alpha \tag{5-113}$$

$$\frac{\partial \psi}{\partial y} = g \cos \alpha \tag{5-114}$$

and

$$\frac{\partial}{\partial x}\left[\rho \frac{\partial p}{\partial x} + \rho^2 g \sin \alpha\right] = \frac{\phi \mu}{K} \frac{\partial \rho}{\partial t} \tag{5-115}$$

since

$$\frac{\partial p}{\partial y} = \frac{\partial \rho}{\partial y} = 0 \tag{5-116}$$

is assumed.

We also have for an ideal liquid

$$\rho \frac{\partial p}{\partial x} = \frac{1}{c} \frac{\partial \rho}{\partial x} \tag{5-117}$$

and thus

$$\frac{\partial}{\partial x}\left[\frac{\partial \rho}{\partial x} + \rho^2 gc \sin \alpha\right] = \frac{\phi \mu c}{K} \frac{\partial \rho}{\partial t} \tag{5-118}$$

Note that in this equation x is positive updip. If we take $x' = -x$ to be positive downdip (as in the paper by Howard and Rachford)[9] we have

$$\frac{\partial}{\partial x'}\left[\frac{\partial \rho}{\partial x'} - \rho^2 gc \sin \alpha\right] = \frac{\phi \mu c}{K} \frac{\partial \rho}{\partial t} \tag{5-119}$$

The important point here is that this differential equation is not linear in the dependent variable, ρ. Hence the usual analytical means of solution are not applicable. Howard and Rachford have solved this equation by numerical methods and shown that several approximate methods for including the effects of gravity yield results in rather close agreement with the exact solution.

Here we show how an approximate solution to equation (5-119) can be obtained.

With ψ defined by

$$\psi = \int_{p_0}^{p} \frac{dp}{\rho(p)} - gx' \sin \alpha \tag{5-120}$$

for x' positive downdip and $y = 0$, we have

$$\frac{\partial}{\partial x'} \left[\rho^2 \frac{\partial \psi}{\partial x'} \right] = \frac{\phi\mu}{K} \frac{\partial\rho}{\partial t} \tag{5-121}$$

as another form of the flow equation. Then noting that

$$\frac{\partial\rho}{\partial t} = c\rho \frac{\partial p}{\partial t} = c\rho^2 \frac{\partial\psi}{\partial t} \tag{5-122}$$

we have

$$\frac{\partial}{\partial x'} \left[\rho^2 \frac{\partial \psi}{\partial x'} \right] = \frac{\phi\mu c}{K} \rho^2 \frac{\partial\psi}{\partial t} \tag{5-123}$$

Thus

$$\rho^2 \frac{\partial^2\psi}{\partial x'^2} + 2\rho \frac{\partial\psi}{\partial x'} \frac{\partial\rho}{\partial x'} = \frac{\phi\mu c}{K} \rho^2 \frac{\partial\psi}{\partial t} \tag{5-124}$$

But from equation (5-120) and the equation of state

$$\frac{\partial\rho}{\partial x'} = c\rho^2 \left(\frac{\partial\psi}{\partial x'} + g \sin \alpha \right) \tag{5-125}$$

and hence

$$\frac{\partial^2\psi}{\partial x'^2} + 2c\rho \frac{\partial\psi}{\partial x'} \left(\frac{\partial\psi}{\partial x'} + g \sin \alpha \right) = \frac{\phi\mu c}{K} \frac{\partial\psi}{\partial t} \tag{5-126}$$

This shows that if the fluid is only very slightly compressible and the gradient of potential is not too large then ψ satisfies

$$\frac{\partial^2\psi}{\partial x'^2} = \frac{\phi\mu c}{K} \frac{\partial\psi}{\partial t} \tag{5-127}$$

to a high degree of approximation.

Initially, at $t = 0$, the system is in hydrostatic equilibrium.

Thus

$$-\frac{K}{\mu}\rho\frac{\partial\psi}{\partial x'} = 0 \tag{5-128}$$

at $t = 0$ for all x', and hence ψ is a constant. We denote the initial value by

$$\psi(x, 0) = \psi_0 = \text{constant} \tag{5-129}$$

Then introducing the new variable

$$\Delta\psi = \psi_0 - \psi(x, t) \tag{5-130}$$

we have

$$\frac{\partial^2\Delta\psi}{\partial x'^2} = \frac{\phi\mu c}{K}\frac{\partial\Delta\psi}{\partial t} \tag{5-131}$$

and $\Delta\psi$ is zero at $t = 0$.

This can now be solved by the same methods as employed for the linear horizontal aquifer. Imposing a constant drawdown at $x' = 0$ is equivalent to fixing a constant value of $\Delta\psi$, say $(\Delta\psi)_\sigma$ at $x' = 0$.

The Laplace transform of the solution for constant drawdown at $x' = 0$ is

$$\overline{\Delta\psi} = (\Delta\psi)_\sigma \exp\left(-x'\sqrt{\frac{\phi\mu c s}{K}}\right) \tag{5-132}$$

Thus the Laplace transform of the gradient of $\Delta\psi$ is

$$\frac{\overline{\partial\Delta\psi}}{\partial x'} = -\frac{\overline{\partial\psi}}{\partial x'} = (\Delta\psi)_\sigma \sqrt{\frac{\phi\mu c}{K}}\sqrt{s}\exp\left(-x'\sqrt{\frac{\phi\mu c s}{K}}\right) \tag{5-133}$$

and hence the flow rate at $x' = 0$ has the transform

$$\bar{q}_\sigma = \frac{K\rho_\sigma A}{\mu}(\Delta\psi)_\sigma\sqrt{\frac{\phi\mu c s}{K}} \tag{5-134}$$

where A is the area and ρ_σ is the value of ρ at $x' = 0$. From this, we obtain

$$Q_\sigma(t) = (\Delta\psi)_\sigma\rho_\sigma A\sqrt{\frac{\phi c K}{\pi\mu}}\sqrt{t} \tag{5-135}$$

which is exactly the same as the cumulative influx from the horizontal aquifer except that $\rho_\sigma(\Delta\psi)_\sigma$ replaces $p_0 - p_\sigma$. For very small compressibility

$$\rho_\sigma(\Delta\psi)_\sigma \approx p_0 - p_\sigma \tag{5-136}$$

and hence for sufficiently small c and also small $p_0 - p_\sigma$ the tilt of the aquifer does not affect the cumulative influx.

The above analysis shows that the distribution of ψ in the tilted aquifer is essentially the same as in the horizontal aquifer, but the pressure distributions are very different.

5.60: Flow of An Ideal Gas, Numerical Integration

All the preceding discussions of transient-flow problems and methods of solution have been concerned with compressible liquids. Many problems of practical importance in the chemical industries and in the production of natural gas can be considered in terms of the flow of ideal gases through porous media. These problems are of a type very different from those of liquid flow.

The differential equation describing the flow of an ideal gas can be written as outlined in Chapter 3 to include the Klinkenberg effect. We obtain with Darcy's law, the ideal gas law and the equation of continuity

$$\frac{\partial}{\partial x_1}\left[p\,\frac{K_\infty}{\mu}\left(1+\frac{b}{p}\right)\frac{\partial p}{\partial x_1}\right]+\frac{\partial}{\partial x_2}\left[p\,\frac{K_\infty}{\mu}\left(1+\frac{b}{p}\right)\frac{\partial p}{\partial x_2}\right]=\phi\,\frac{\partial p}{\partial t} \tag{5-137}$$

for flow in two dimensions.

This equation can be put in a more useful form by introducing

$$P=\frac{p+b}{p_i+b}$$

$$\theta=\frac{K_\infty(p_i+b)t}{2\phi\mu L^2} \tag{5-138}$$

$$\bar{x}_1=\frac{x}{L},\qquad \bar{x}_2=\frac{x_2}{L}$$

where p_i is the initial uniform pressure in the system and L is a characteristic dimension of the system. In terms of these variables the differential equation takes the form

$$\frac{\partial^2 P^2}{\partial \bar{x}_1{}^2}+\frac{\partial^2 P^2}{\partial \bar{x}_2{}^2}=\frac{\partial P}{\partial \theta} \tag{5-139}$$

This is a non-linear second-order partial differential equation and hence is not amenable to solution by any of the conventional analytical methods. Approximate solutions can be obtained by writing the equation in the form

$$\frac{\partial^2 P^2}{\partial \bar{x}_1{}^2}+\frac{\partial^2 P^2}{\partial \bar{x}_2{}^2}=\frac{1}{2P}\,\frac{\partial P^2}{\partial \theta} \tag{5-140}$$

and observing that if the variations in P are small compared to the average value of P then the coefficient of $\partial P^2/\partial\theta$ on the right can be treated as a constant.

Thus if over the whole region of flow, and for all values of θ, P differs only slightly from its initial value, then

$$\frac{\partial^2 P^2}{\partial \bar{x}_1{}^2} + \frac{\partial^2 P^2}{\partial \bar{x}_2{}^2} \approx \frac{1}{2}\frac{\partial P^2}{\partial\theta} \tag{5-141}$$

or, in terms of the original variables

$$\frac{\partial^2(p+b)^2}{\partial x_1{}^2} + \frac{\partial^2(p+b)^2}{\partial x_2{}^2} = \frac{\phi\mu}{K_\infty(p_i+b)}\frac{\partial(p+b)^2}{\partial t} \tag{5-142}$$

This is similar in form to the corresponding equation for a compressible liquid.

The equation in exact form, equation (5-133) can be solved by numerical integration by employing the method of finite differences. Such solutions have been carried out by Jenkins and Aronofsky[1, 12] and by Bruce, Peaceman and Rachford.[3] This method of solution is described for the case of linear flow in a system of length L. One end is closed to flow while at the other the pressure is suddenly reduced and held constant. Thus the boundary and initial conditions are

$$P(\bar{x}, 0) = 1$$

$$\frac{\partial P^2}{\partial \bar{x}}(1, \theta) = 0 \tag{5-143}$$

$$P(0, \theta) = H = \frac{p_0 + b}{p_i + b}$$

Here p_0 is the constant value of p at the opened end of the system.

Most of the numerical results reported in the forementioned papers are reported for the case of $b = 0$; that is, neglecting the Klinkenberg effect. However, the mathematical formulation and, indeed, the numerical results are the same whether b is considered zero or not, provided one considers the results expressed in terms of the dimensionless variables defined above.[6]

The procedure for solving the equation

$$\frac{\partial^2 P^2}{\partial \bar{x}^2} = \frac{\partial P}{\partial\theta} \tag{5-144}$$

consists of replacing $\bar{x}$ and θ by the discrete variables,

$$x_j = j\Delta\bar{x} \quad \text{and} \quad \theta_n = n\Delta\theta,$$

where j and n are integers and replacing the derivatives by finite differences.

Thus, if we expand P^2 in a Taylor series, we obtain

$$P^2_{j+1,n} = P^2_{j,n} + \left(\frac{\partial P^2}{\partial \bar{x}}\right)_{j,n} \Delta\bar{x} + \frac{1}{2!}\left(\frac{\partial^2 P^2}{\partial \bar{x}^2}\right)_{j,n} (\Delta\bar{x})^2 + \dots . \tag{5-145}$$

and similarly

$$P^2_{j-1,n} = P^2_{j,n} - \left(\frac{\partial P^2}{\partial \bar{x}}\right)_{j,n} \Delta\bar{x} + \frac{1}{2!}\left(\frac{\partial^2 P^2}{\partial \bar{x}^2}\right)_{j,n} (\Delta\bar{x})^2 + \dots . \tag{5-146}$$

where $P_{j,n}$ denotes the value of $P(x_j, \theta_n)$. Adding these two equations yields

$$\left(\frac{\partial^2 P^2}{\partial \bar{x}^2}\right)_{j,n} = \frac{P^2_{j+1,n} + P^2_{j-1,n} - 2P^2_{j,n}}{(\Delta\bar{x})^2} + 0[(\Delta\bar{x})^2] \tag{5-147}$$

where $0[(\Delta\bar{x})^2]$ denotes a term of the order of $(\Delta\bar{x})^2$.

A similar expansion for $P_{j,n+1}$ in terms of $\Delta\theta$ yields

$$\left(\frac{\partial P}{\partial \theta}\right)_{j,n} = \frac{P_{j,n+1} - P_{j,n}}{\Delta\theta} + 0(\Delta\theta) \tag{5-148}$$

Then, neglecting terms of the order of $\Delta\theta$ and $(\Delta\bar{x})^2$, yields

$$P_{j,n+1} = P_{j,n} + \frac{\Delta\theta}{(\Delta\bar{x}^2)} [P^2_{j+1,n} + P^2_{j-1,n} - 2P^2_{j,n}] \tag{5-149}$$

as the difference equation approximating to the differential equation, equation (5-144). This is called the predictive form of the difference equation since $P_{j,n+1}$ is given in terms of quantities evaluated at $\theta = \theta_n$.

This equation must be supplemented with the initial conditions

$$P_{j,0} = 1, \qquad\qquad 0 < j < N \tag{5-150}$$

where

$$N\Delta\bar{x} = 1 \tag{5-151}$$

and the boundary conditions

$$P_{0,n} = H \tag{5-152}$$

and

$$\frac{P^2_{N+1,n} - P^2_{N-1,n}}{\Delta\bar{x}} = 0 \tag{5-153}$$

This last condition introduces an image point at $\bar{x} = (N + 1)\Delta\bar{x}$ to assure no flow at the closed end. The value of P^2 at this point is never used since equation (5-153) is just used to eliminate $P^2_{N+1,n}$ from the difference equation (5-149), at $j = N$.

The integration begins by using (5-149) to compute $P_{j,i}$ from the initial values, making use of the boundary conditions.

This procedure is limited by errors. Thus, certain errors are introduced by the truncation of the Taylor series in setting up the difference equations. Also errors are introduced by round off in the calculations. These errors may either accumulate or decay as the difference equation is applied iteratively for succeeding time steps. It can be shown that errors will not accumulate if

$$\frac{(\Delta \bar{x})^2}{\Delta \theta} \geq 4P_n \tag{5-154}$$

Here P_n denotes the average of the $P_{j,n}$ at the nth time step.

Obviously, the truncation errors are made small by making both $\Delta \theta$ and $\Delta \bar{x}$ small but here a further restriction is imposed. This stability condition, equation (5-154), must be satisfied and imposes a limitation on the size of time steps which can be used.

Other forms of the difference equation can be constructed which are less severely limited than the one above. Thus Bruce *et al* used[3]

$$P_{j,n+1} - \frac{\Delta \theta}{2(\Delta \bar{x})^2} \Delta_{j,n+1}^2 P^2 = P_{j,n} + \frac{\Delta \theta}{2(\Delta \bar{x}^2)} \Delta_{j,n}^2 P^2 \tag{5-155}$$

where

$$\Delta_{j,n}^2 P^2 = P_{j+1,n}^2 + P_{j-1,n}^2 - 2 P_{j,n}^2 \tag{5-156}$$

as the difference representation of the differential equation. This equation is of implicit form, that is, $P_{j,n+1}$ cannot be expressed explicitly in terms of values at the nth time level. Special methods must be used in this case; however the stability requirement

$$\frac{(\Delta \bar{x})^2}{\Delta \theta} > 2(P_n - P_{n-1}) \tag{5-157}$$

is much less restrictive. The reader is referred to the literature cited above for details of this method.

The results of the numerical integrations obtained by the workers cited above are exemplified in Figures 5-4 and 5-5. Figure 5-4 illustrates the results obtained by Aronofsky[1] for the linear case including the Klinkenberg effect. Here the system of length L and initial pressure p_i has the end at $x = 0$ opened to the constant pressure, p_0, at time $t = 0$. Observe that as the parameter $H = (p_0 + b)/(p_i + b)$ is nearer to unity the solution approaches the liquid solution.

The radial case of a well at the center of a closed circular reservoir is shown in Figure 5-5. Here the gas is withdrawn at a constant mass rate.

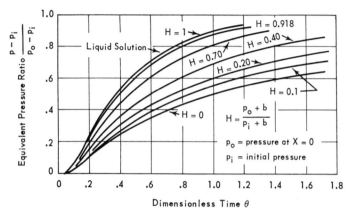

Figure 5–4. Flow of an ideal gas in a linear system showing the pressure variation with time at the closed end of a tube when the pressure at the other end is suddenly reduced. (*After Aronofsky, 1954.*)

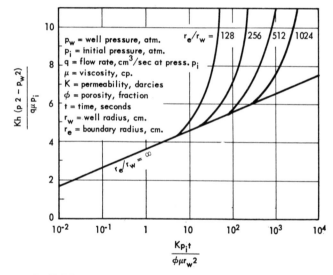

Figure 5–5. Radial flow of an ideal gas; circular well at center of circular reservoir producing at constant mass rate. (*After Jenkens and Aronofsky, 1953.*)

Observe that in terms of $p^2 - p_0^2$ the solution is very similar to the liquid case shown in Figure 5-3.

EXERCISES

1. Use the superposition of point sinks in the plane to show that shutting in one well and producing another yields pressure information which can be used to deduce properties of an oil reservoir. (This is a pressure interference test.)

2. Use superposition of point sinks in the plane to deduce the form of the pressure build-up curve for a shut-in test following a brief period of production after an earlier shut-in test. Consider the flow rate constant prior to the first shut-in and a different constant following the first shut-in.

3. Show that for $b = 0$

$$\frac{p^2 - p_i^2}{2} = -\frac{q_i p_i \mu}{4\pi Kh}\left\{-Ei\left(-\frac{\phi\mu r^2}{4Kp_i t}\right)\right\}$$

with q_i = volume production rate measured at pressure p_i and $r^2 = x_1^2 + x_2^2$, is a solution of equation (5-135), and hence the pressure build-up for a well in an infinite uniform gas reservoir is approximated by

$$\frac{p_w^2 - p_i^2}{2} \approx \frac{q_i p_i \mu}{4\pi Kh}\ln\frac{\delta t}{t_s + \delta t}, \, \delta t \gg 0$$

References

1. Aronofsky, J. S., Jenkins, R., *Proc. 1st. U. S. Natl. Cong. Appl. Mech.*, 763, Chicago (1951).
2. Aronfsky, J. S., *J. Appl. Phy.*, **25**, 48 (1954).
3. Bruce, G. H., Peaceman, D. W., Rachford, H. H., Rice, J. P., *Trans. AIME* **198**, 79 (1953).
4. Carslaw, H. S., and Jaeger, J. C., "Conduction of Heat in Solids," 2nd Ed., Oxford Univ. Press, London (1959).
5. Churchill, R. V., "Modern Operational Mathematics in Engineering," McGraw-Hill Book Co., New York (1944).
6. Collins, R. E., Crawford, P. B., *Trans. AIME*, **198**, 339 (1953).
7. Hicks, A. L., Weber, A. G., Ledbetter, R. L., *Trans. AIME*, **216**, 400 (1959).
8. Horner, D. R., *Proc. 3rd World Pet ol. Congr.* Sect. II, E. J. Brill, Leiden (1951).
9. Howard, D. S., Jr., Rachford, H. H., Jr., *Trans. AIME*, **207**, 92 (1956).
10. Hurst, W., *Trans. AIME*, **151**, 57 (1943).
11. Hurst, W., Van Everdingen, A. F., *Trans. AIME*, **186**, 305 (1949).
12. Jenkins, R., Aronofsky, J. S., *J. Appl. Mech.*, **20**, 210 (1953).
13. Magnus, W., Oberhettinger, F., "Formulas and Theorems for the Special Functions of Mathematical Physics," Springer Verlag, Berlin (1943), Chelsea Publishing Co., New York (1949).
14. Matthews, C. S., Brous, F., Hazebrock, P., *Trans. AIME*, **201**, 182 (1954).
15. Nisle, R. B., *Trans. AIME*, **213**, 85 (1958).
16. Stegemeier, G. L., Matthews, C. S., *Trans. AIME*, **213**, 44 (1958).
17. Van Everdingen, A. F., *Trans. AIME*, **198**, 171 (1953).

6. SIMULTANEOUS FLOW OF IMMISCIBLE FLUIDS

6.10: Steady Linear Flow and the Measurement of Relative Permeabilities; the Boundary Effect

The concept of relative permeability as described in Chapter 3 is fundamental to the study of the simultaneous flow of immiscible fluids through porous media. However, the determination of relative permeability as a function of saturation in the laboratory is not so simple as it might seem at first sight. This becomes evident when we consider the mathematical description of the phenomenon of two-phase flow.

We consider a linear system of length L initially containing a wetting fluid at irreducible saturation, S_c (connate water saturation), and a non-wetting fluid at saturation, $1 - S_c$. A mixture of the two fluids is introduced uniformly over one end of the system, $x = 0$, and removed at the other end. Both fluids are considered incompressible and the input rates of the two fluids are to be held constant. This process brings about changes of saturation within the system, but if the input is maintained indefinitely the outflow composition will eventually become the same as the inflow composition. The saturation distribution within the system ceases to change with time. Steady linear flow then exists.

For steady flow the equation of continuity applied to each of the two incompressible fluids yields

$$\frac{\partial v_w}{\partial x} = 0 \tag{6-1}$$

and

$$\frac{\partial v_{nw}}{\partial x} = 0 \tag{6-2}$$

And by Darcy's law

$$v_w = -\frac{K_w}{\mu} \frac{\partial p_w}{\partial x} = \text{constant} \tag{6-3}$$

and

$$v_{nw} = -\frac{K_{nw}}{\mu_{nw}} \frac{\partial p_{nw}}{\partial x} = \text{constant} \tag{6-4}$$

139

Here the pressures in the two fluids are related by

$$p_{nw} - p_w = p_c \qquad (6\text{-}5)$$

where p_c is the capillary pressure. In this case the imbibition capillary-pressure curve applies.

If the saturations, S_w and S_{nw}, are uniform within the medium then K_w, K_{nw} and p_c are independent of x. Then equations (6-3) and (6-4) can be integrated to yield

$$K_w = \frac{\mu_w q_w L}{A \Delta p_w} \qquad (6\text{-}6)$$

and

$$K_{nw} = \frac{\mu_{nw} q_{nw} L}{A \Delta p_{nw}} \qquad (6\text{-}7)$$

Here q_w and q_{nw} are the volumetric flow rates of the two fluids, A is the cross-sectional area of the sample and Δp_w and Δp_{nw} are the respective pressure differences across the system. Since p_c is uniform, these pressure differences must be equal.

In actual fact, if the flow rates are sufficiently high the assumption of a uniform saturation distribution is a reasonably good approximation. When such is the case measurements of q_w, q_{nw}, Δp and the saturation of the sample for a variety of values of q_w/q_{nw} are sufficient to define the permeabilities as functions of saturation. For low flow rates, however, the assumption of uniform saturation is not valid.

It is observed that at the outflow face of a system such as this the wetting fluid will not flow out until the saturation at the outflow face has built up to some critical value. The reason for this behavior lies in the capillary capillary pressure-saturation relationship.

Continuity of fluid pressure must exist everywhere within each continuous fluid. Initially, only non-wetting fluid exists exterior to the outflow face. Zero capillary pressure exists exterior to the outflow face. Consequently, no wetting fluid may appear outside this face until the capillary pressure is reduced to zero at the outflow face. Since wetting-fluid saturation is increasing here, the imbibition capillary pressure curve applies. On this curve capillary pressure goes to zero at a critical value of wetting-fluid saturation, denoted here by $1 - S_{ro}$. That is, S_{ro} is the critical value of non-wetting fluid saturation. This point can be seen on the curve in Figure 2-4 in Chapter 2.

At the critical saturation of non-wetting fluid, S_{ro}, the non-wetting fluid forms a discontinuous phase within the porous medium. Non-wetting fluid then ceases to flow for any finite pressure gradient. In the case of water-oil systems in water-wet porous media, this is called the residual oil saturation. At $S_{nw} = S_{ro}$, the permeability to non-wetting fluid, K_{nw}, is therefore zero.

For any saturation of wetting fluid below the critical saturation, at the outflow face, only non-wetting fluid will flow. However, at the critical saturation the permeability to the non-wetting phase becomes zero. Thus an infinite pressure gradient must exist in the non-wetting fluid at the outflow face for this critical saturation if flow of non-wetting fluid is to occur.

From equations (6-3), (6-4) and (6-5) and the definitions of q_w and q_{nw}

$$q_{nw} = -\frac{K_{nw}A}{\mu_{nw}}\frac{\partial p_w}{\partial x} - \frac{K_{nw}A}{\mu_{nw}}\frac{dp_c}{dS_w}\frac{\partial S_w}{\partial x} \qquad (6\text{-}8)$$

and

$$q_w = -\frac{K_wA}{\mu_w}\frac{\partial p_w}{\partial x} \qquad (6\text{-}9)$$

At the critical saturation, denoted by $S_{nw} = S_{ro}$ or $S_w = 1 - S_{ro}$, $K_w \approx K$, $K_{nw} = 0$ and dp_c/dS_w is either zero or a finite negative value. Then, since $q_{nw} \neq 0$, it follows that for $S_{nw} \to S_{ro}$ at $x = L$ we must have

$$\lim_{x \to L}\left(\frac{\partial S_w}{\partial x}\right) = +\infty \qquad (6\text{-}10)$$

This phenomenon is called the outlet boundary effect, or the end effect. Boundary effects also occur at the inlet face. As soon as the input fluid mixture is exposed to the inflow face the wetting fluid begins to imbibe into the porous medium due to the action of capillary forces. In the steady state this tends to disappear. Thus in the steady state $\partial S_w/\partial x$ is zero at the inflow face.

For steady flow $\partial S_w/\partial x$ will be zero everywhere except for a small region near the outflow face. The extent of this region of non-uniform saturation decreases as the flow rate increases. Thus if a steady-state technique is employed for the measurement of relative permeability, high rates must be used or some means of correcting for the end effect must be used. Various methods of measuring relative permeabilities are reviewed and compared by Richardson et al.[11]

The distribution of saturation existing during the steady-flow regime considered here can be computed. $\partial p_w/\partial x$ can be eliminated between equations (6-8) and (6-9) and an equation involving only $\partial S_w/\partial x$ and functions of S_w obtained, i.e. K_w, K_{nw} and p_c. For fixed values of q_w and q_{nw}, this equa-

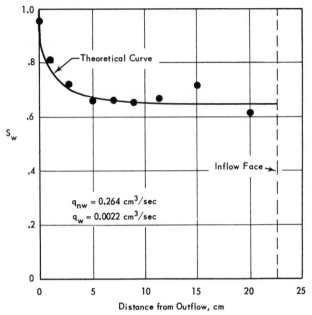

Figure 6–1. Saturation distribution in a linear system under steady-state conditions; illustration of outlet boundary effect. (*After Richardson et al., 1952.*)

tion can be integrated by graphical or numerical means if K_w, K_{nw} and p_c are known as functions of S_w.

Such calculations have been compared with experiment by Richardson et al.[1] A typical curve showing the agreement between theory and experimental data is shown in Figure 6-1.

6.20: Immiscible Displacement; The Buckley-Leverett Equation

The displacement of oil from a porous medium by water plays an important role in the production of petroleum. In both natural-water drive and secondary-water flooding this displacement process is fundamental. For the case of one-dimensional flow of incompressible immiscible fluids, a simple mathematical description of this process can be formulated, provided the interfacial tension between the fluids is small so that capillary-pressure effects can be neglected. It is also necessary to neglect gravity effects. This is described as follows.

We consider first the case of linear displacement in a thin tube of porous material inclined at an angle α to the horizontal. Distance x measured along the tube is considered positive going up the tube. The tube has cross-sectional area A which is small enough to consider the pressure and saturation uniform in any cross section.

From Darcy's law we have for the two phases, wetting and non-wetting

$$q_w = -\frac{K_w A}{\mu_w} \left[\frac{\partial p_w}{\partial x} + \rho_w g \sin \alpha \right] \tag{6-11}$$

and

$$q_{nw} = -\frac{K_{nw} A}{\mu_{nw}} \left[\frac{\partial p_{nw}}{\partial x} + \rho_{nw} g \sin \alpha \right] \tag{6-12}$$

where

$$p_{nw} - p_w = p_c \tag{6-13}$$

defines capillary pressure.

The fluids are considered to be incompressible so that the continuity equation applied to each phase yields the two equations

$$\frac{\partial q_w}{\partial x} = -\phi A \frac{\partial S_w}{\partial t} \tag{6-14}$$

and

$$\frac{\partial q_{nw}}{\partial x} = -\phi A \frac{\partial S_{nw}}{\partial t} \tag{6-15}$$

where the saturations are related by

$$S_w + S_{nw} = 1 \tag{6-16}$$

Adding equations (6-14) and (6-15) yields, in view of equation (6-16)

$$\frac{\partial}{\partial x} (q_w + q_{nw}) = 0 \tag{6-17}$$

so that the total flow rate

$$q = q_w + q_{nw} \tag{6-18}$$

is a constant along the tube.

In view of this result, we can define the fraction, f_w, of the flowing stream which is wetting fluid by the equation

$$f_w = \frac{q_w}{q} \tag{6-19}$$

and similarly for the non-wetting fluid

$$f_{nw} = \frac{q_{nw}}{q} = 1 - f_w \tag{6-20}$$

Then the continuity equations read

$$\frac{q}{\phi A}\frac{\partial f_w}{\partial x} = -\frac{\partial S_w}{\partial t} \tag{6-21}$$

and

$$\frac{q}{\phi A}\frac{\partial f_{nw}}{\partial x} = -\frac{\partial S_{nw}}{\partial t} \tag{6-22}$$

Now if we combine equations (6-11), (6-12) and (6-13) to eliminate p_w and p_{nw} we obtain

$$q_{nw} = -\frac{K_{nw}}{\mu_{nw}} A \left[-\frac{\mu_w q_w}{K_w A} + \frac{\partial p_c}{\partial x} - \Delta \rho g \sin \alpha \right] \tag{6-23}$$

where

$$\Delta \rho = \rho_w - \rho_{nw} \tag{6-24}$$

Then the substitutions

$$q_w = f_w q \tag{6-25}$$

and

$$q_{nw} = (1 - f_w)q \tag{6-26}$$

yield

$$f_w = \frac{1 + \dfrac{K_{nw}A}{\mu_{nw}q}\left[\dfrac{\partial p_c}{\partial x} - \Delta \rho g \sin \alpha\right]}{1 + \dfrac{K_{nw}\mu_w}{K_w \mu_{nw}}} \tag{6-27}$$

for f_w .

If the total flow rate, q, is very high and/or interfacial tension and density difference are small, then this can be approximated by

$$f_w \approx \left(1 + \frac{K_{nw}\mu_w}{K_w \mu_{nw}}\right)^{-1} = f_w\left(S_w , \frac{\mu_w}{\mu_{nw}}\right) \tag{6-28}$$

which is a function of saturation only, the viscosity ratio being just a parameter. When this approximation is valid

$$\frac{\partial f_w}{\partial x} = \frac{df_w}{dS_w}\frac{\partial S_w}{\partial x} \tag{6-29}$$

and the continuity equation then reads

$$\left[\frac{q}{\phi A} \frac{df_w}{dS_w} \right] \frac{\partial S_w}{\partial x} = -\frac{\partial S_w}{\partial t} \qquad (6\text{-}30)$$

Note that the continuity equation for the other phase reduces to this same equation.

This equation is a non-linear equation in S_w since the coefficient of $\partial S_w / \partial x$ is a function of S_w ; hence solutions cannot be obtained by classical analytical means. However, a numerical technique can be employed to determine the saturation distribution as follows.

The total derivative of $S_w(x, t)$ with respect to time is

$$\frac{dS_w}{dt} = \frac{\partial S_w}{\partial x} \frac{dx}{dt} + \frac{\partial S_w}{\partial t} . \qquad (6\text{-}31)$$

Thus, if $x = x(t)$ is chosen to coincide with a surface of fixed S_w

$$\frac{dS_w}{dt} = 0 \qquad (6\text{-}32)$$

and

$$\left(\frac{dx}{dt} \right)_{S_w} = -\frac{\partial S_w}{\partial t} \left(\frac{\partial S_w}{\partial x} \right)^{-1} \qquad (6\text{-}33)$$

as the rate of advance of the saturation S_w .

If we use equation (6-30) to eliminate $\partial S_w / \partial t$ from this equation, we obtain

$$\left(\frac{dx}{dt} \right)_{S_w} = \frac{q}{\phi A} \frac{df_w(S_w)}{dS_w} \qquad (6\text{-}34)$$

This equation is called the Buckley-Leverett equation.[2] Integrating with respect to t there results

$$x_{S_w}(t) - x_{S_w}(0) = \frac{Q(t) - Q(0)}{\phi A} \frac{df_w(S_w)}{dS_w} \qquad (6\text{-}35)$$

Here $x_{S_w}(t)$ and $x_{S_w}(0)$ are the coordinates of the plane at which the saturation S_w exists at time t and time zero, respectively, and $Q(t)$ and $Q(0)$ are the cumulative total volumes passed through the system at times t and zero, respectively.

Since the coefficient, df_w/dS_w , can be evaluated for every saturation, S_w , if the permeability ratio, K_{nw}/K_w is known as a function of saturation, it follows that when the saturation distribution is known at $t = 0$, the

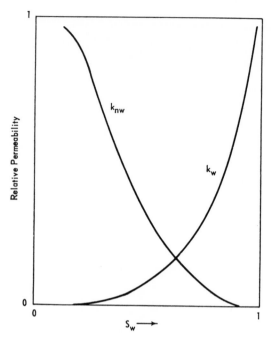

Figure 6-2. Relative permeability curves.

saturation distribution at any time $t > 0$ can be computed from this distribution and equation (6-35).

However, certain difficulties are encountered in this process. This is best illustrated by example. Consider the relative permeability curves shown in Figure 6-2.* From these the curve of f_w versus S_w for $\mu_w/\mu_{nw} = 1$ shown in Figure 6-3 is constructed. Also, we can evaluate df_w/dS_w as a function of S_w as shown in this same figure.

Now we consider an initial saturation distribution consisting of all saturations, $S_c < S_w < 1$ existing at $x = 0$ at time $t = 0$. Applying equation (6-35) to this distribution we obtain the saturation distribution shown in

---

* Note that since only ratios of permeability are to be used

$$\frac{k_w}{k_{nw}} = \frac{K_w}{K_{nw}}$$

where

$$k_w = \frac{K_w}{K}, \quad k_{nw} = \frac{K_{nw}}{K}$$

and K is the single-phase permeability of the medium.

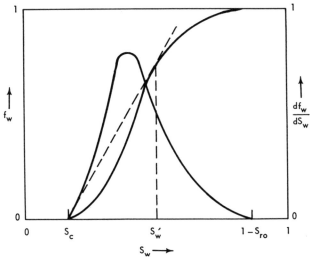

Figure 6–3. Fraction of wetting fluid in flowing fluid as a function of wetting-fluid saturation; also the first-derivative curve showing the determination of saturation at the displacement front.

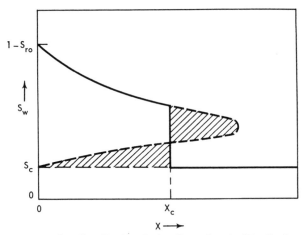

Figure 6–4. Saturation distribution during linear immiscible displacement as computed from the Buckley-Leverett equation showing the discontinuity in saturation as required by a material balance.

Figure 6-4 at a later time $t > 0$. Here the curve of df_w/dS_w in Figure 6-2 has been used. Observe that multiple values of saturation are developed. This is physically impossible.

To eliminate the multiple values of saturation account is taken of the fact that at $t = 0$, $S_w = S_c$ everywhere and wetting fluid is injected at $x = 0$, i.e. $f_w = 1$ at $x = 0$. Thus a volumetric balance for wetting fluid yields

$$Q(t) = \int_0^{x_c} \phi(S_w - S_c)A \ dx \tag{6-36}$$

Here x_c is a cutoff point beyond which $S_w = S_c$. Thus we are introducing a discontinuity in saturation at $x = x_c$. This discontinuity can also be studied from the standpoint of shock formation.[12]

Integrating this integral by parts, there results

$$Q = \phi A x_c[S_w' - S_c] - \int_{1-S_{ro}}^{S_w'} \phi A x \ dS_w \tag{6-37}$$

where S_w' is the upper saturation at the discontinuity. In view of $Q(0) = 0$ and $x_{S_w}(0) = 0$ in equation (6-35), this can be written

$$Q = \phi A[S_w' - S_c]x_c - Q \int_{1-S_{ro}}^{S_w'} \frac{df_w}{dS_w} \ dS_w \tag{6-38}$$

and integration then yields with $f_w = 1$ at $S_w = 1 - S_{ro}$

$$Q = \phi A[S_w' - S_c]x_c - Q[f_w(S_w') - 1] \tag{6-39}$$

or

$$\phi A x_c = \frac{f_w(S_w')}{S_w' - S_c} Q \tag{6-40}$$

But again using equation (6-35) with our conditions on $Q(0)$ and $x(0)$ gives

$$\phi A x_c = Q \frac{df_w}{dS_w} (S_w') \tag{6-41}$$

and hence

$$\frac{df_w(S_w')}{dS_w} = \frac{f_w(S_w')}{S_w' - S_c} \tag{6-42}$$

This shows that the saturation distribution remains single valued if all saturations below S_w' are removed. This critical, or cutoff saturation, is determined as the value of S_w at which the straight line passing through the point $f_w = 0$, $S_w = S_c$ is tangent to the curve $f_w(S_w)$. This line is shown in Figure 6-3. The tangent point determining S_w' is also shown.

The discontinuity in saturation introduced by this procedure is shown in Figure 6-4. Further analysis shows that the areas enclosed in the two shaded loops are equal.

This procedure for analyzing the displacement process is widely used in the petroleum industry. Many elaborations and extensions are possible. For example, the same analysis carried through for a horizontal system of axial symmetry, radial flow, yields the basic equation corresponding to equation (6-35) as

$$\phi \pi h[r_{S_w}^2(t) - r_{S_w}^2(0)] = [Q(t) - Q(0)] \frac{df_w(S_w)}{dS_w} \tag{6-43}$$

Here $r_{S_w}(t)$ is the radius at which S_w exists at time t, and h is the thickness of the stratum. The cutoff saturation S_w' is still given by equation (6-42).

It should be noted that the saturation, $S_w = 1 - S_{ro}$, is not propagated along the system. Consequently, $S_w < 1 - S_{ro}$ everywhere within the system. This should be anticipated because K_{nw} becomes zero at the critical saturation, $S_{nw} = S_{ro}$ (the residual oil saturation). Although the relative permeability curves are for all practical purposes independent of the nature of the fluids this residual saturation is not purely a characteristic of the porous medium. Some experimental evidence indicates that lowering of interfacial tension between the fluids may produce a reduction of the residual saturation of non-wetting fluid.[6]

6.21: The Welge Integration of the Buckley-Leverett Equation; The Calculation of Relative Permeabilities from Displacement Data

When capillary and gravity effects are negligible the displacement of one fluid by another in a linear flow system can be described by the Buckley-Leverett equation provided, of course, that the fluids are immiscible and incompressible and the porous medium is homogeneous.

Consider a linear displacement with $S_w(x, 0) = S_c$ but now stipulate that the system is of length L. Wetting fluid is introduced to displace non-wetting fluid at rate $q(t)$ at $x = 0$ for $t > 0$. Thus at $x = 0$, only wetting fluid is flowing while at $x = L$ both fluids are flowing. The cumulative production of non-wetting fluid from the outflow end is denoted by Q_{nw} and the cumulative inflow of wetting fluid by Q. Note that Q is equivalent to the total outflow, $Q_w + Q_{nw}$.

We have now at the outflow end,

$$f_{nw} = \frac{q_{nw}}{q} = \frac{dQ_{nw}}{dQ} = \left(1 + \frac{K_w \mu_{nw}}{K_{nw} \mu_w}\right)^{-1} \tag{6-44}$$

from the definition of f_{nw}. Thus

$$\frac{K_{nw}}{K_w} = \frac{\mu_{nw}}{\mu_w} \left(1 - \frac{dQ_{nw}}{dQ}\right)^{-1} \frac{dQ_{nw}}{dQ} \tag{6-45}$$

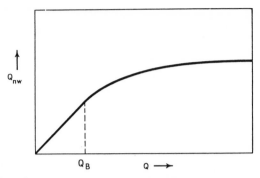

Figure 6–5. Cumulative recovery of non-wetting fluid versus cumulative volume of injected wetting fluid in a linear immiscible displacement experiment.

Hence if for such a linear displacement experiment Q_{nw} and Q are measured, then at any stage of the displacement the ratio of permeabilities can be estimated from an estimate of dQ_{nw}/dQ. However, the saturation corresponding to this ratio is the instantaneous value at the outflow face and this is not directly measurable. This problem is overcome by the technique of Welge.[14]

A material balance on the wetting fluid yields

$$\phi A \int_0^L S_{nw} \, dx = \phi A L(1 - S_c) - Q_{nw} \tag{6-46}$$

Then integrating the left member by parts yields

$$\phi A L S_{nw}(L) - \phi A \int_{S_{ro}}^{S_{nw}(L)} x \, dS_{nw} = \phi A L(1 - S_c) - Q_{nw} \tag{6-47}$$

Observing that $dS_w = -dS_{nw}$ and using

$$\phi A x_{S_w} = Q \frac{df_w(S_w)}{dS_w} \tag{6-48}$$

there results

$$S_{nw}(L) = \frac{1}{\phi A L} \left[\phi A L(1 - S_c) - Q_{nw} + Q \frac{dQ_{nw}}{dQ} \right] \tag{6-49}$$

Here use has been made of equation (6-44).

All quantities on the right side of this equation can be determined from the data. To illustrate one method of applying these results, consider the data shown in Figure 6-5 from a typical displacement experiment. These data represent the linear waterflooding ($\mu_w = 1$) of a sample initially containing oil of viscosity $\mu_{nw} = 6$ at a saturation, $S_{nw} = 0.80$.*

* It should be noted that the shape of the curve, Q_{nw} versus Q, is strongly dependent on the viscosity ratio as well as on the relative permeability curve.

Observe that from $Q = 0$ up to a value $Q = Q_B$ the curve of Q_{nw} versus Q is a straight line with slope one. Thus only oil is produced up to this point. This point is termed water breakthrough. At water breakthrough, the saturation discontinuity which is the leading boundary of the watered region reaches the outflow end of the system.

For $Q > Q_B$, the curve Q_{nw} versus Q can be represented by

$$Q_{nw} = a + b \ln Q \tag{6-50}$$

over any *small* segment. Here a and b are constants, for any small segment of the curve, to be determined from the data. That is, we use a number of data points on a segment and the method of least squares to determine a value of b for the segment. Then

$$Q \frac{dQ_{nw}}{dQ} = b \tag{6-51}$$

for this segment. This can then be used in equations (6-45) and (6-49) along with the value of Q_{nw} to compute the permeability ratio and $S_{nw}(L)$. In this manner the curve shown in Figure 6-6 is obtained. Recently an extension of the Welge technique has been devised which permits evaluation of the individual fluid permeabilities from displacement data of this type. This procedure is due to Johnson *et al.*[7]

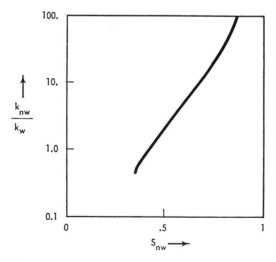

Figure 6-6. Relative permeability ratio computed from a linear displacement experiment.

6.30: Linear Immiscible Displacement Including the Effects of Capillary Pressure

The description of linear immiscible displacement provided by the Buckley-Leverett equation is a good approximation to the real physical situation only for very high flow rates. At lower flow rates the effects of capillary pressure cannot be ignored.

For two- or three-dimensional two-phase flow of immiscible fluids the mathematical description takes the form of two simultaneous second-order partial differential equations, but for the one-dimensional case these can be reduced to a single second-order partial differential equation. Thus for the linear case we have the continuity equations for the two phases, equations (6-14) and (6-15), with the supplementary equations (6-16) and (6-18). The fluids are considered incompressible.

Darcy's law for each phase is as given in equations (6-11) and (6-12) with the supplementary equation, equation (6-13), defining capillary pressure. These equations combined with equation (6-18),

$$q = q_{nw} + q_w$$

yield

$$\frac{\partial p_w}{\partial x} = -\frac{q}{A\left(\dfrac{K_w}{\mu_w} + \dfrac{K_{nw}}{\mu_{nw}}\right)} - \frac{\dfrac{dp_c}{dS_{nw}}}{1 + \dfrac{K_w\mu_{nw}}{K_{nw}\mu_w}}\frac{\partial S_{nw}}{dx} + \frac{\dfrac{K_w\rho_w}{\mu_w} + \dfrac{K_{nw}\rho_{nw}}{\mu_{nw}}}{\dfrac{K_w}{\mu_w} + \dfrac{K_{nw}}{\mu_{nw}}}\,g\sin\alpha. \quad (6\text{-}52)$$

Here

$$\frac{\partial p_c}{\partial x} = \frac{dp_c}{dS_{nw}}\frac{\partial S_{nw}}{\partial x} \quad (6\text{-}53)$$

has been assumed. This is valid if the fluids and porous medium are each homogeneous. From this, there results by Darcy's law

$$q_w = \frac{q}{1 + \dfrac{K_{nw}\mu_w}{K_w\mu_{nw}}} + \frac{A\dfrac{K_w}{\mu_w}\dfrac{dp_c}{dS_{nw}}}{1 + \dfrac{K_w\mu_{nw}}{K_{nw}\mu_w}}\frac{\partial S_{nw}}{\partial x} \quad (6\text{-}54)$$

if the tube of flow is horizontal.

Then since q is independent of x this expression, when substituted into the continuity equation for the wetting fluid, yields

$$\frac{\partial}{\partial x}\left[\frac{q}{1 + \dfrac{K_{nw}\mu_w}{K_w\mu_{nw}}} + \frac{A\dfrac{K_w}{\mu_w}\dfrac{dp_c}{dS_{nw}}}{1 + \dfrac{K_w\mu_{nw}}{K_{nw}\mu_w}}\frac{\partial S_{nw}}{\partial x}\right] = \phi A\frac{\partial S_{nw}}{\partial t} \quad (6\text{-}55)$$

which is the partial differential equation in the dependent variable S_{nw}. This equation is non-linear in S_{nw} and, therefore, cannot be solved by conventional analytical methods. Before considering numerical methods of solution, we will write the equation in a more concise but general form as follows.

Define the dimensionless variables

$$\zeta = \frac{x}{L}, \quad \lambda = \frac{qt}{\phi A L} \tag{6-56}$$

where L is the length of the system. Also define a dimensionless capillary pressure as

$$\bar{p}_c = \frac{p_c(S_{nw})}{\left(\dfrac{dp_c}{dS_{nw}}\right)_{\text{char.}}} \tag{6-57}$$

Here "char." denotes a characteristic value. Note that S_{nw} itself is already a dimensionless quantity.

In terms of these quantities the differential equation takes the form

$$\frac{\partial}{\partial \zeta}\left[g(S)\frac{\partial S}{\partial \zeta}\right] + h(S)\frac{\partial S}{\partial \zeta} = \frac{\partial S}{\partial \lambda} \tag{6-58}$$

where

$$S = S_{nw} \tag{6 59}$$

$$g(S) = \frac{1}{B}\frac{k_w k_{nw}\dfrac{d\bar{p}_c}{dS}}{k_w + k_{nw}\dfrac{\mu_w}{\mu_{nw}}} \tag{6-60}$$

and

$$h(S) = \frac{d}{dS}\left(\frac{1}{1 + \dfrac{k_{nw}\mu_w}{k_w\mu_{nw}}}\right) \tag{6-61}$$

In these equations the relative permeabilities have been introduced and

$$B = \frac{qLu_{nw}}{AK\left(\dfrac{dp_c}{dS}\right)_{\text{char.}}} \tag{6-62}$$

is a dimensionless constant parameter; K is the single-phase permeability of the medium.

Equation (6-58) describes the saturation of the non-wetting fluid as a function of the dimensionless variables ζ and λ, and the dimensionless parameter B. However, for purposes of numerical solution, it is more convenient to define another dependent variable as a function of S. Thus, let S_c denote the minimum wetting-fluid saturation (connate water saturation), and let S_{ro} denote the minimum non-wetting fluid saturation obtainable by displacement (residual oil saturation). Then define

$$r(s) = \frac{1}{Z} \int_{1-S_c}^{S} g(\eta) \, d\eta \qquad (6\text{-}63)$$

where

$$Z = \int_{1-S_c}^{S_{ro}} g(\eta) \, d\eta \qquad (6\text{-}64)$$

In terms of $r(s)$, the differential equation becomes

$$\frac{\partial^2 r}{\partial \zeta^2} + \alpha(r) \frac{\partial r}{\partial \zeta} = \beta(r) \frac{\partial r}{\partial \lambda} \qquad (6\text{-}65)$$

where

$$\alpha(r) = \frac{1}{Z} \frac{d}{dr} \left(\frac{1}{1 + \dfrac{k_{nw}\mu_w}{k_w\mu_{nw}}} \right) \qquad (6\text{-}66)$$

and

$$\beta(r) = \frac{1}{Z} \frac{dS}{dr} \qquad (6\text{-}67)$$

Before the solution of the problem can be attempted we must supplement the differential equation with appropriate boundary and initial conditions. At $t = 0$, let

$$S(x, 0) = 1 - S_c \qquad (6\text{-}68)$$

or, in terms of $r(\zeta, \lambda)$

$$r(\zeta, 0) = 0 \qquad (6\text{-}69)$$

Letting $x = 0$ be the inflow boundary we have, since only wetting fluid is being injected

$$q_{nw}(0, t) = 0 \qquad (6\text{-}70)$$

This implies that

$$\frac{\partial p_{nw}}{\partial x} = 0 \qquad (6\text{-}71)$$

and/or

$$q = q_w \tag{6-72}$$

at $x = 0$. Thus, from equation (6-5) and Darcy's law,

$$\frac{-\partial p_w}{\partial \xi} = \frac{\partial p_c}{\partial \zeta} = \frac{dp_c}{dS} \frac{\partial S}{\partial \zeta} = \frac{Lq\mu_w}{AK_w}, \quad \text{at} \quad x = 0 \tag{6-73}$$

This is equivalent to

$$\frac{\partial r(0, \lambda)}{\partial \zeta} = \frac{1}{Z} \left[1 + \frac{k_w \mu_{nw}}{k_{nw} \mu_w} \right]^{-1} = v[r(0, \lambda)] \tag{6-74}$$

Here v denotes the function of $r(0, \lambda)$ on the left.

At the outflow face the boundary effect must be considered. No wetting fluid is produced until the saturation of wetting fluid at the outflow face has built up to its maximum value, $1 - S_{ro}$. From then on, the saturation of wetting fluid remains at this maximum value. Thus if $t = t^*$ is the time at which S_w first reaches the maximum value: at $x = L$

$$q_w(L, t) = 0, \quad t < t^* \tag{6-75}$$

and

$$S(L, t) = S_{ro}, \quad t > t^* \tag{6-76}$$

In terms $r(\zeta, \lambda)$, these can be shown to be

$$\frac{\partial r(1, \lambda)}{\partial \zeta} = -\frac{1}{Z} \left[1 + \frac{k_{nw} \mu_w}{k_w \mu_{nw}} \right]^{-1} = y[r(1, \lambda)], \quad \lambda < \lambda^* \tag{6-77}$$

$y(r)$ denoting the function of r on the left, and

$$r(1, \lambda) = 1, \quad \lambda > \lambda^* \tag{6-78}$$

where $\lambda^* = qt^*/LA\phi$.

The formulation of the problem is now complete. This formulation is essentially that given by Douglas, Blair and Wagner.[3] The numerical method of solution to be described now is also due to these investigators.

Let $\Delta\zeta = 1/N$ and $\Delta\lambda > 0$, and let $\xi_i = (i - 1)\Delta\zeta$, $i = 1, 2, \cdots N + 2$ and $\lambda_n = n\Delta\lambda$, n being an integer. Denote by $W_{i,n}$ the value of the solution of the difference equation (the approximation to r) at the point (ζ_i, λ_n). Define the first- and second-difference quotients by

$$\Delta_\zeta W_{i,n} = \frac{W_{i+1,n} - W_{i-1,n}}{2\Delta\zeta} \tag{6-79}$$

and

$$\Delta_{\zeta}^{2}W_{i,n} = \frac{W_{i+1,n} + W_{i-1,n} - 2W_{i,n}}{(\Delta\zeta)^2} \tag{6-80}$$

Then approximate the second derivative, $\partial^2 r/\partial\zeta^2$, by

$$\frac{\partial^2 r}{\partial\zeta^2} \approx \frac{1}{2}\,[\Delta_{\zeta}^{2}W_{i,n+1} + \Delta_{\zeta}^{2}W_{i,n}] \tag{6-81}$$

and similarly

$$\frac{\partial r}{\partial\zeta} \approx \frac{1}{2}\,[\Delta_{\zeta}W_{i,n+1} + \Delta_{\zeta}W_{i,n}] \tag{6-82}$$

Also by approximation

$$\frac{\partial r}{\partial\lambda} \approx \frac{W_{i,n+1} - W_{i,n}}{\Delta\lambda} \tag{6-83}$$

Now with respect to the dimensionless time variable λ the above approximations are centered at the level $n + \frac{1}{2}$. Thus it is desired to evaluate the coefficients in the differential equation at a value of $W \approx r$ at this level. This value of W is approximated by

$$W^{*}_{i,n+1/2} = W_{i,n} + \frac{\Delta\lambda[\Delta_{\zeta}^{2}W_{i,n} + \alpha(W_{i,n})\Delta_{\zeta}W_{i,n}]}{2\beta(W_{i,n})} \tag{6-84}$$

which is itself obtained by an obvious difference approximation to the differential equation.

The initial and boundary conditions must be introduced also in the difference formulation. Thus the initial condition is

$$W_{i,0} = 0 \tag{6-85}$$

The boundary conditions need to be evaluated at the time level $n + 1$, thus let

$$W^{**}_{i,n+1} = W_{i,n} + \frac{\Delta\lambda[\Delta_{\zeta}^{2}W_{i,n} + \alpha(W_{i,n})\Delta_{\zeta}W_{i,n}]}{\beta(W_{i,n})} \tag{6-86}$$

which is similar to the way in which $W^{*}_{i,n+1/2}$ was expressed.

Then the boundary conditions can be expressed as

$$\Delta_{\zeta}W_{1,\,n+1} = v(W^{**}_{1,\,n+1})$$

$$\Delta_{\zeta}W_{N+1,\,n+1} = y(W^{**}_{N+1,\,n+1}), \; \lambda < \lambda^*. \tag{6-87}$$

$$W_{N+1,\,n+1} = 1, \; \lambda > \lambda^*$$

These, together with the difference analog of the differential equation

$$\frac{1}{2}\,\Delta_\zeta^2 W_{i,n+1} + \Delta_\zeta W_{i,n}] + \alpha(W^*_{i,n+1/2})\,\frac{1}{2}\,[\Delta_\zeta W_{i,n+1} + \Delta_\zeta W_{i,n}]$$

$$= \beta\;(W^*_{i,n+1/2})\,\frac{W_{i,n+1} - W_{i,n}}{\Delta\lambda}$$

(6-88)

and the definitions of $W^*_{i,n+1/2}$ and $W^{**}_{i,n+1}$ form a system of algebraic equations for the $W_{i,n}$ which can be solved by a form of Gaussian elimination.[3] Of course, a high-speed electronic digital computer is required for such calculations.

Douglas, Blair and Wagner[3] have obtained numerical solutions by this method for a variety of values of the dimensionless flow rate parameter B, equation (6-62). All of their calculations were for a viscosity ratio, $\mu_w/\mu_{nw} = \frac{1}{2}$, and a particular set of relative permeability and capillary-pressure curves. These curves are shown in Figure 6–7. The equations for these curves are

$$k_{nw} = 1.425(s - 0.216)$$

$$k_w = 1.6329(0.7 - s)^2$$

$$\bar{p}_t = \left[\frac{0.05669}{(0.7 - s)^2} - 0.242\right]$$

(6-89)

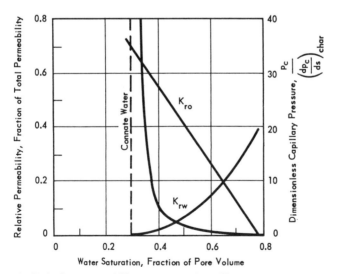

Figure 6–7. Relative permeability curves and capillary pressure curves for calculation of linear immiscible displacement including the effects of capillary pressure. (*After Douglas et al., 1958.*)

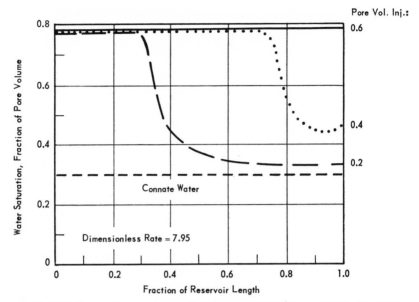

Figure 6–8. Saturation versus fraction of length during linear immiscible displacement showing effects of capillary pressure. (*After Douglas et al., 1958.*)

From these and the definition of $r(s)$, the polynomial representations

$$S = 0.7 - 0.43372 \ r^3 + 3.08195 \ r^6 \ - 10.592 \ r^9 - 14.256 \ r^{12} - 6.69618 \ r^{15}$$

$$\frac{k_w}{k_w + \tfrac{1}{2}k_{nw}} = [0.6291r^3 - 3.4696r^6 + 12.229r^9 - 16.178r^{12} + 7.7895r^{15}]^2 \qquad (6\text{-}90)$$

are obtained.*

Typical results of these calculations are shown in Figures 6–8 and 6–9. Figure 6-8 shows the wetting-fluid saturation plotted versus ζ for several values of λ. Note that ζ is distance expressed as a fraction of total length and λ is number of pore volumes injected. Figure 6-9 shows the wetting-fluid saturation plotted versus ζ at $\lambda = 0.2$ pore volumes injected for several values of the rate parameter, B. Also shown is the distribution of saturation as computed by the Buckley-Leverett equation. This corresponds to an infinite value of B.

These curves show quite clearly that for very high rates the Buckley-Leverett equation gives a relatively good approximation to the actual saturation distribution.

---

* Certain limitations are inherent in these representations. The reader is referred to the paper by Douglas *et al.* for a discussion of these.

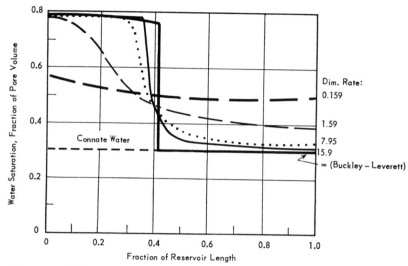

Figure 6–9. Effect of rate on saturation distribution during linear immiscible displacement at 0.2 pore volume injected. (*After Douglas et al., 1958.*)

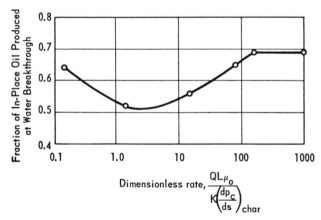

Figure 6–10. Computed correlation between recovery at breakthrough and dimensionless rate parameter for linear immiscible displacement. (*After Douglas et al., 1958.*)

The effect of the dimensionless rate factor on the production of non-wetting fluid from the system is shown very clearly in Figure 6-10. Here the fraction of total in-place non-wetting fluid produced at the time of first wetting-fluid production is plotted versus B. This curve was obtained from their calculations by Douglas *et al.*

Several important features on this curve are to be noted, first, the plateau for large values of B. This is expected since for very high rates the Buckley-Leverett equation applies and recovery becomes independent of rate. At very low rates the entire process becomes one of imbibition rather than displacement and recovery is high. The striking feature is that a minimum exists on the curve. This is a real effect which is to be seen in systems of this type.

Kyte and Rappoport[9] have reported data from experimental studies of water displacing oil from water-wet porous media which have the same character as the computed results of Douglas *et al.* These data were obtained with a core holder of the type shown in Figure 6-11.

Typical data as obtained by Kyte and Rappoport are shown in Figure 6-12. These data are for a viscosity ratio $\mu_w/\mu_{nw} = 0.621$. Neither the relative permeability nor capillary-pressure curves were reported. The permeabilities of the alundum cores used in the experiments were between 508 md. and 574 md. Since (dp_c/dS_{nw}) char. has been omitted from the rate factor on the horizontal scale exact comparison to the computed results is not possible. Also note that here μ_w instead of μ_{nw} appears in the group on the horizontal scale.

In spite of the limitations of these data it is quite evident that the general character of the computed results is in agreement with similar results from experiments. Such confirmation substantiates the entire mathematical theory of immiscible flow based on Darcy's law.

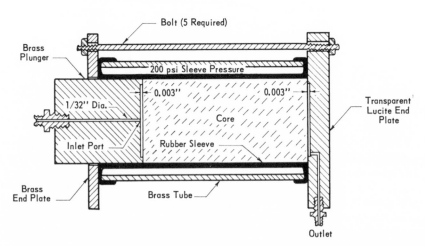

Figure 6–11. Core holder for linear immiscible displacement experiments. (*After Kyte and Rappoport, 1958.*)

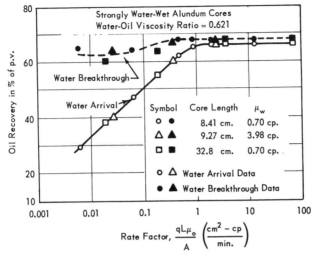

Figure 6–12. Typical experimental data showing effect of rate on recovery for linear immiscible displacement. (*After Kyte and Rappoport, 1958.*)

6.40: Imbibition

In the study of linear displacement treated in the previous section the important role of capillarity in determining the character of the displacement was emphasized. It is possible for one fluid to replace another in a porous medium by capillary forces alone. This process is termed imbibition.

The macroscopic description of the imbibition process can be formulated most easily for a linear system. This brings out all the basic characteristics of the phenomenon.

Consider a cylindrical sample of a homogeneous porous material of length L and cross-sectional area A. Let the sample have impermeable boundaries on the lateral surface and one end. Consider the medium to have an initial wetting fluid saturation, S_{wi}, and a non-wetting fluid saturation, $1 - S_{wi}$.

If the unsealed face is now exposed to wetting fluid this fluid will imbibe into the porous medium spontaneously due to capillary forces, with a resultant countercurrent flow of non-wetting fluid.*

Since for incompressible fluids the total volume of fluid in the sample is constant, it follows that

$$q = q_w + q_{nw} = 0 \tag{6-91}$$

at any cross section.

* Note that this is much the same process as occurs in an ordinary ink blotter.

From Darcy's law, we have

$$q_w = -\frac{K_w A}{\mu_w} \frac{\partial p_w}{\partial x} \tag{6-92}$$

and

$$q_{nw} = -\frac{K_{nw} A}{\mu_{nw}} \frac{\partial p_{nw}}{\partial x} \tag{6-93}$$

where

$$p_{nw} - p_w = p_c \tag{6-94}$$

defines the capillary pressure. In these equations x is taken positive into the imbibition face. These equations then yield in view of equation (6-91)

$$\frac{\partial p_w}{\partial x} = \frac{-\dfrac{K_{nw}}{\mu_{nw}}}{\dfrac{K_{nw}}{\mu_{nw}} + \dfrac{K_w}{\mu_w}} \frac{dp_c}{dS_{nw}} \frac{\partial S_{nw}}{\partial x} \tag{6-95}$$

Then it follows that

$$q_w = \frac{K_w K_{nw} A}{K_w \mu_{nw} + K_{nw} \mu_w} \frac{dp_c}{dS_{nw}} \frac{\partial S_{nw}}{\partial x} \tag{6-96}$$

and

$$q_{nw} = -q_w \tag{6-97}$$

These equations can be combined with the continuity equations for the two fluids. Either equation can be used since $S_w + S_{nw} = 1$ makes these dependent. Thus we obtain from

$$\frac{\partial q_{nw}}{\partial x} = -\phi A \frac{\partial S_w}{\partial t} \tag{6-98}$$

the equation

$$\frac{\partial}{\partial x}\left[\frac{K_w K_{nw}}{K_w \mu_{nw} + K_{nw}\mu_w} \frac{dp_c}{dS_{nw}} \frac{\partial S_{nw}}{\partial x} \right] = \phi \frac{\partial S_{nw}}{\partial t} \tag{6-99}$$

as the differential equation determining the saturation as a function of x and t. Note that we have neglected gravity effects here. Also note that this equation is the same as equation (6·55) of the previous section with $q = 0$. This is as it should be since the same basic mechanisms are involved.

The boundary conditions follow from physical considerations. At the closed boundary, $x = L$, no flow of either fluid occurs. This is imposed by requiring zero pressure gradient in both fluids at $x = L$. In view of equations (6-92), (6-93) and (6-94) this is equivalent to zero gradient of capillary pressure or hence zero gradient of saturation. Thus at $x = L$

$$\left(\frac{\partial S_{nw}}{\partial x}\right)_{x=L} = 0 \tag{6-100}$$

At the inflow face we should require continuity of capillary pressure. Since p_c is zero in the free fluid and this value occurs at $S_{nw} = S_{ro}$ in the porous medium we require

$$S_{nw}(0, t) = S_{ro} \tag{6-101}$$

at the inflow face. These boundary conditions coupled with the initial condition, $S_{nw} = 1 - S_{wi}$, and the differential equation are sufficient to determine the saturation distribution as a function of x and t.

Blair[1] has formulated the imbibition problem in a different way. From Darcy's law for each fluid and the two continuity equations, he writes

$$\left.\begin{aligned} \frac{\partial}{\partial x}\left(M\frac{\partial P}{\partial x}\right) + \frac{\partial}{\partial x}\left(N\frac{\partial V}{\partial x}\right) &= 0 \\[2mm] \frac{\partial}{\partial x}\left(N\frac{\partial P}{\partial x}\right) + \frac{\partial}{\partial x}\left(M\frac{\partial V}{\partial x}\right) &= -4\phi\frac{dS_w}{dp_c}\frac{\partial V}{\partial t} \end{aligned}\right\} \tag{6-102}$$

with boundary conditions

$$\left.\begin{aligned} P(0, t) = V(0, t) &= 0 \\[2mm] \frac{\partial P(L, t)}{\partial x} = \frac{\partial V(L, t)}{\partial x} &= 0 \end{aligned}\right\} \tag{6-103}$$

where

$$\left.\begin{aligned} M &= \frac{K_{nw}}{\mu_{nw}} + \frac{K_w}{\mu_w} \\[2mm] N &= \frac{K_{nw}}{\mu_{nw}} - \frac{K_w}{\mu_w} \\[2mm] P &= \tfrac{1}{2}(p_{nw} + p_w) \\[2mm] V &= \tfrac{1}{2}(p_{nw} - p_w) = \tfrac{1}{2}p_c \end{aligned}\right\} \tag{6-104}$$

These equations are then solved by methods of finite differences on a digital computer. The advantage of this formulation lies in the fact that not only p_c, and hence S_{nw} is determined, but also p_{nw} and p_w are determined as functions of x and t. Blair also treats the problem of imbibition into a right circular cylinder.

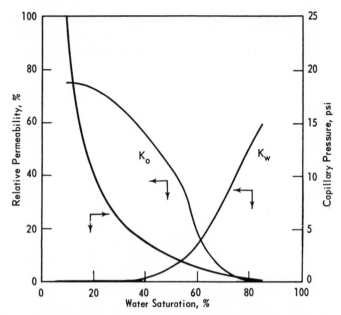

Figure 6–13. Relative permeability and capillary pressure curves. (*After Blair, 1960.*)

Typical results as obtained by Blair are shown in Figure 6-14. These curves correspond to the relative permeability curves and the capillary pressure curve shown in Figure 6-13. Other pertinent data are:

$$\mu_0 = 5 \text{ cp}$$

$$\mu_w = 1 \text{ cp}$$

$$S_{wi} = 9.2\%$$

$$\phi = 32.1\%$$

$$L = 30.48 \text{ cm}$$

$$A = 7.92 \text{ cm}^2$$

$$\text{Elapsed Time} = 6.6 \text{ hr}$$

$$\text{In-Place Oil Prod.} = 36\%$$

$$K = 200 \text{ md}$$

These calculations describe oil production by imbibition of water. Figure 6-15 shows the effect of oil viscosity on the rate of oil production by water imbibition.

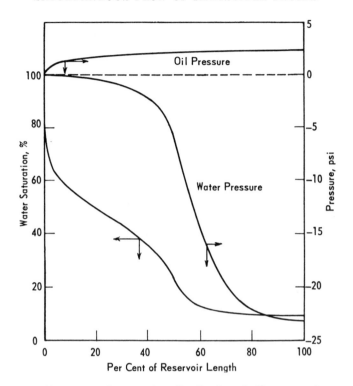

Figure 6–14. Pressure and saturation distributions in linear countercurrent imbibition. (*After Blair, 1960.*)

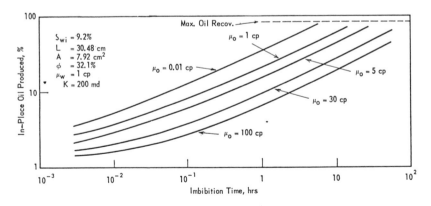

Figure 6–15. Effect of oil viscosity on oil recovery by linear countercurrent imbibition. (*After Blair, 1960.*)

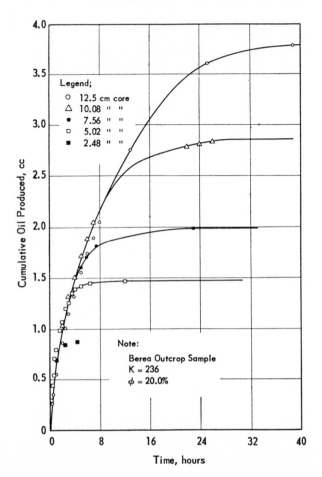

Figure 6–16. Oil recovery versus time for linear countercurrent imbibition showing the effect of system length. (*After Graham and Richardson, 1960.*)

The results obtained by Blair are in good agreement with experimental observations. One point not brought out in Blair's paper is the effect of sample length on the rate of oil production. Figure 6-16 shows typical imbibition results for a natural sandstone which illustrate the effect of sample length. These data are due to Graham and Richardson.[5]

A great deal of insight into the effect of various parameters can be gained by expressing all equations and results in terms of dimensionless quantities. This is discussed in the chapter on models.

6.50: Multi-dimensional Two-phase Flow: Numerical Solutions

In previous sections various problems of linear two-phase flow were treated by several methods. Of all these only one formulation can be extended to the multi-dimensional case.

For flow of two incompressible, immiscible fluids in three dimensions Darcy's law for the two fluids can be written in the form

$$\left. \begin{aligned} \hat{v}_{nw} &= -\frac{K_{nw}}{\mu_{nw}} \nabla \psi'_{nw} \\ \hat{v}_{w} &= -\frac{K_{w}}{\mu_{w}} \nabla \psi_{w}' \end{aligned} \right\} \tag{6-105}$$

Here

$$\left. \begin{aligned} \psi'_{nw} &= p_{nw} + \rho_{nw} g x_3 \\ \psi_{w}' &= p_{w} + \rho_{w} g x_3 \end{aligned} \right\} \tag{6-106}$$

where

$$p_{nw} - p_{w} = p_c \tag{6-107}$$

as usual. ρ_{nw} and ρ_{w} are the respective fluid densities and x_3 is taken as positive vertically.

The continuity equations are in this case.

$$\nabla \cdot \hat{v}_{nw} = -\phi \frac{\partial S_{nw}}{\partial t} \tag{6-108}$$

and

$$\nabla \cdot \hat{v}_{w} = -\phi \frac{\partial S_{w}}{\partial t} \tag{6-109}$$

As in Blair's treatment of imbibition Douglas, Peaceman and Rachford[4] define the variables

$$\left. \begin{aligned} P &= \tfrac{1}{2} (\psi'_{nw} + \psi_{w}') \\ V &= \tfrac{1}{2} (\psi'_{nw} - \psi_{w}') \end{aligned} \right\} \tag{6-110}$$

and also

$$\left. \begin{aligned} M &= \frac{K_{nw}}{\mu_{nw}} + \frac{K_{w}}{\mu_{w}} \\ N &= \frac{K_{nw}}{\mu_{nw}} - \frac{K_{w}}{\mu_{w}} \end{aligned} \right\} \tag{6-111}$$

Then the differential equations of the problem can be written as

$$\nabla \cdot (M \nabla P) + \nabla \cdot (N \nabla V) = 0$$

$$\nabla \cdot (N \nabla P) + \nabla \cdot (M \nabla V) = -4\phi \frac{dS_w}{dp_c} \frac{\partial V}{\partial t} \Bigg\}$$ (6-112)

These equations, or other forms, must be solved by numerical methods. Douglas, Peaceman and Rachford have solved several problems by such methods.

The importance of such solutions to flow problems of this type is that details of the saturation distributions can be determined in this way which could not be determined by experimental studies.

6.60: Turbulent Flow of Immiscible Fluids

Throughout this chapter only the laminar flow regime has been considered. This is the flow regime obtaining in petroleum reservoirs. However, many important problems in the process industries involve the turbulent flow of immiscible fluids through porous media, particularly counter-current flow in vertical packed columns.

The theory of turbulent immiscible flow is not so well developed as for the laminar flow regime. Lerner and Grove[10] reviewed the theoretical and experimental results available up to 1951. A more recent study in which relative permeabilities were measured for turbulent flow conditions has been reported by Stewart and Owens.[13]

EXERCISES

1. Verify equation (6-43) as the correct form of the Buckley-Leverett equation for radial flow.
2. Linear displacement of a non-wetting fluid by a wetting fluid yields data of Q_{nw} versus Q. Here Q_{nw} = cumulative non-wetting fluid produced, Q = cumulative wetting fluid injected. Show that the slope α of the curve, Q_{nw} versus Q, is such that for very large values of μ_{nw}/μ_w, $\alpha \approx 1$ and for very small values of μ_{nw}/μ_w, $\alpha \approx 0$, following breakthrough.
3. The saturation S_w' at the front in a Buckley-Leverett type displacement is given by equation (6-42). This is also the saturation at the outflow end of a linear system at the time of breakthrough of wetting fluid. Show that equation (6-42) can be written as

$$\frac{1}{S_w' - S_c} = -\left[\frac{d}{dS_w} \left(1 + \frac{\mu_w K_{nw}}{\mu_{nw} K_w} \right) \right]_{S_w = S_w'}$$

and deduce the dependence of S_w' on the viscosity ratio.
4. Show that equations (6-102), (6-103) and (6-104) yield the same formulation for the dependence of S_{nw} on x and t as equations (6-98), (6-99), (6-100), and (6-101) during the imbibition process. i.e., these last equations can be deduced from equations (6-102), (6-103) and (6-104).

References

1. Blair, P. M., Paper No. 1475G, AIME Secondary Rec. Symp., Wichita Falls, Texas, May 1960.
2. Buckley, S. E., and Leverett, M. C., *Trans AIME*, **146,** 107 (1942).
3. Douglas, J., Jr., Blair, P. M., and Wagner, R. J., *Trans. AIME*, **215,** 96 (1958).
4. Douglas, J., Jr., Peaceman, D. W., and Rachford, H. H., Jr., *Trans.* AIME, **216,** 297 (1959).
5. Graham, J. W., and Richardson, J. G. (Private communication).
6. Guerraro, E. T., and Kennedy, H. T., *Trans. AIME*, **201,** 124 (1954).
7. Johnson, E. F., Gassler, D. P., and Naumann, V. O., *Trans. AIME*, **216,** 370 (1959).
8. Jordan, J. K., McCardell, W. M., and Hocott, C. R., *Oil Gas J.*, **55,** (19) 98 (1957).
9. Kyte, J. R., and Rappoport, L. A., *Trans. AIME*, **215,** 423 (1958).
10. Lerner, B. J., and Grove, C. S., *Ind. Eng. Chem.*, **43,** 216 (1951).
11. Richardson, J. G., Kerver, J. K., Hafford, J. A., and Osoba, J., *Trans. AIME*, **195,** 187 (1952).
12. Sheldon, J. W., Zondek, B., and Cardwell, W. T. Jr., *Trans. AIME*, **216,** 290 (1959).
13. Stewert, C. R., and Owens, W. W., *Trans. AIME*, **215,** 121 (1958).
14. Welge, H., *Trans. AIME*, **195,** 91 (1952).

7. MOVING BOUNDARY PROBLEMS; DISPLACEMENT; DEPOSITION OF SOLIDS

7.10: Displacement of One Fluid by Another

The displacement of one fluid by another within a porous medium can frequently be represented as a moving boundary problem. Thus suppose that a porous medium is saturated with oil of viscosity μ_o and density ρ_o, except for a small amount of immobile connate water of viscosity μ_w and density ρ_w. Denoting the connate water saturation by S_c, the oil saturation is $1 - S_c$.

Now suppose that water is injected to displace the oil, and assume that the region in which appreciable saturation gradients exist is so small that the porous medium can be divided into two parts: one region containing only oil and immobile connate water and another region containing only water and immobile residual oil. The first region is ahead of the "front" and the other behind the "front." The surface separating these two regions is a mathematical surface representing a discontinuity in saturation. Ahead of this front only oil is mobile and behind only water is mobile. Such a front corresponds to the saturation discontinuity introduced with the Buckley-Leverett equation in section 6-20. For certain viscosity ratios, or unusual relative permeability characteristics the regions formed approximate those described here.

Before considering the complete mathematical description of a displacement problem of this type, we examine the boundary conditions which must apply on the boundary.

Consider a small cylinder of length δs and cross-sectional area δA with its axis perpendicular to the front as indicated in Figure 7-1. At time t let the front coincide with the end of the cylinder. At a later time $t + \delta t$ the front will have moved a distance

$$\delta s = v_f \delta t \tag{7-1}$$

normal to the front. Here v_f is the rate of advance, or velocity of the front in the direction of the normal to the front.

During this time a volume of water entered the cylinder. This volume is given by

$$v_{wn} \delta A \delta t$$

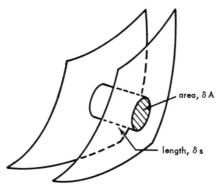

Figure 7–1. Displacement front and tube of flow.

where v_{wn} is the rate of flow of water per unit area normal to the front at this point. This volume of water must be equal to the volume of water now filling the cylinder in the length δs except for the connate water originally present. Thus

$$v_{wn}\delta A\,\delta t = \delta A\,\delta s\,(1 - S_{ro})\phi - \delta A\,\delta s\,S_c\phi \tag{7-2}$$

where S_{ro} is the residual oil saturation behind the front and ϕ is the porosity of the medium. Equations (7-1) and (7-2) then yield

$$v_f = \frac{v_{wn}}{\phi(1 - S_{ro} - S_c)} \tag{7-3}$$

A similar consideration of the quantity of oil leaving the other end of the cylinder yields

$$v_f = \frac{v_{on}}{\phi(1 - S_{ro} - S_c)} \tag{7-4}$$

where v_{on} is the component of oil flow rate per unit area normal to the boundary. We thus have

$$v_{on} = v_{wn} \tag{7-5}$$

on the moving boundary.

In the region behind the front, the water region, only water is flowing. Here Darcy's law and the equation of continuity yield a partial differential equation for ψ_w, the flow potential for water. Ahead of the front, in the oil region, these yield a partial differential equation for ψ_{nw}, the flow potential for oil. If capillary pressure is neglected, then on the moving boundary these potentials must yield the same fluid pressure but not the same flow potential. This follows from the fact that the fluid pressure must be single valued.

If gravity effects can be neglected, then the problem can be formulated entirely in terms of pressure. In this case the dependent variable in the differential equation is p in both regions but still two solutions are required, one behind the front and a different solution ahead of the front. One exceptional case does exist in which the same function describes the pressure distribution everywhere. (See section 7-20.)

In displacement processes of this kind in which the front can be represented by a mathematical surface the following formulation may be used.

A function $F(x_1, x_2, x_3, t)$ can be employed to represent a two-dimensional surface in the x_1, x_2, x_3 space which moves and changes form in time by noting that

$$F(x_1, x_2, x_3, t) = \alpha = \text{constant} \tag{7-6}$$

for fixed t, describes a one-parameter family of surfaces. That is, a unique surface is specified by assigning to α a definite constant value. In general, changing the value of t while holding α fixed yields another surface. This other surface may be viewed as the original surface which has moved and undergone change in shape during the elapsed time.

Thus, in equation (7-6) x_1, x_2, x_3 are the coordinates of a particular point on the front at time t and

$$F(x_1 + \delta x_1, x_2 + \delta x_2, x_3 + \delta x_3, t + \delta t) = \alpha \tag{7-7}$$

specifies that this same point is located at $x_1 + \delta x_1, x_2 + \delta x_2, x_3 + \delta x_3$ at time $t + \delta t$.

Applying Taylor's expansion to equation (7-7), and taking note of (7-6) yields

$$\frac{\partial F}{\partial x_1} \delta x_1 + \frac{\partial F}{\partial x_2} \delta x_2 + \frac{\partial F}{\partial x_3} \delta x_3 + \frac{\partial F}{\partial t} \delta t = 0 \tag{7-8}$$

to first order in the increments δx_i and δt. Here the derivatives are evaluated at x_1, x_2, x_3, t. Since the point of the front is a fluid particle which moves with a velocity given by

$$\hat{v}_p = \frac{\hat{v}_w}{\phi(1 - S_{ro} - S_c)} \tag{7-9}$$

we can write

$$\delta x_i = \frac{v_{wi}\delta t}{\phi(1 - S_{ro} - S_c)} \tag{7-10}$$

and then

$$v_{w1} \frac{\partial F}{\partial x_1} + v_{w2} \frac{\partial F}{\partial x_2} + v_{w3} \frac{\partial F}{\partial x_3} + \phi(1 - S_{ro} - S_c) \frac{\partial F}{\partial t} = 0 \qquad (7\text{-}11)$$

Of course v_o could replace v_w in this equation.

The solution of this equation describes the evolution of the boundary in time. Observe that in order to solve for F it is necessary that the v_{wi} be known functions. Thus the potential distribution must be either a priori determined or else determined simultaneously with F. In general as the surface moves the potential distribution changes. The only case in which this is not so is that in which the "oil" and "water" are indistinguishable. That is, the densities, viscosities and permeabilities are the same for the two fluids. Thus one may consider "blue" water displacing "red" water. Then the position of the boundary has no effect on the potential distribution; the potential problem may be solved first and then the boundary equation solved.

Problems which can actually be solved in this manner are few. Muskat[8] gives a technique suitable for some problems. An alternative with greater applicability is given in section 7.30 following the simple case below.

7.11: Analytical Solution of the Linear Displacement Problem

Here we consider the linear frontal displacement of oil by water in a homogeneous porous medium. Let the length of the medium be L and assume gravity and capillary effects negligible.

Applying Darcy's law and the equation of continuity, we have for the pressure distribution

$$\frac{\partial^2 p_w}{\partial x^2} = 0, \qquad 0 < x < x_f \qquad (7\text{-}12)$$

$$\frac{\partial^2 p_o}{\partial x^2} = 0, \qquad x_f < x < L \qquad (7\text{-}13)$$

$$\left. \begin{array}{l} p_o = p_w \\[2mm] \dfrac{K_{wr}}{\mu_w} \dfrac{\partial p_w}{\partial x} = \dfrac{K_{oc}}{\mu_o} \dfrac{\partial p_o}{\partial x} \end{array} \right\} x = x_f(t) \qquad \begin{array}{l} (7\text{-}14) \\[4mm] (7\text{-}15) \end{array}$$

$$p_w = p_1, \qquad x = 0 \qquad (7\text{-}16)$$

and

$$p_o = p_2, \qquad x = L \qquad (7\text{-}17)$$

Here K_{wr} is the permeability to water at residual oil saturation and K_{oc} is the permeability to oil at connate water saturation. Equation (7-15) follows from the boundary condition formulated as equation (7-5) in the previous section.

Integrating equations (7-12) and (7-13) yields

$$p_w = Ax + B \qquad (7\text{-}18)$$

and

$$p_o = A'x + B' \qquad (7\text{-}19)$$

Applying the boundary conditions yields

$$B = p_1$$

$$A = -\frac{\Delta p}{mL + (1 - m)x_f}$$

$$A' = -\frac{m\Delta p}{mL + (1 - m)x_f} \qquad (7\text{-}20)$$

$$B' = \frac{-(1 - m)x_f\Delta p}{mL + (1 - m)x_f} + p_1$$

Here

$$m = \frac{K_{wr}\mu_o}{K_{oc}\mu_w} \qquad (7\text{-}21)$$

is called the mobility ratio and

$$\Delta p = p_1 - p_2 \qquad (7\text{-}22)$$

is the applied pressure differential.

From these equations and

$$v_w = -\frac{K_{wr}}{\mu_w}\frac{\partial p_w}{\partial x} = -\frac{K_{wr}}{\mu_w}A \qquad (7\text{-}23)$$

there results the representation of equation (7-3)

$$\frac{dx_f}{dt} = \frac{K_{wr}\Delta p}{\mu_w\phi(1 - S_c - S_{ro})}\frac{1}{mL + (1 - m)x_f} \qquad (7\text{-}24)$$

for the rate of advance of the interface. Integrating this with $x_f = 0$ at $t = 0$ yields

$$t = \frac{\mu_w\phi(1 - S_c - S_{ro})}{K_{wr}\Delta p}[mLx_f + \tfrac{1}{2}(1 - m)x_f{}^2] \qquad (7\text{-}25)$$

for the time required for the front to advance from $x = 0$ to $x = x_f$.

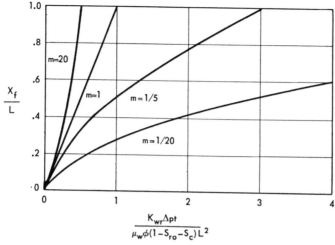

Figure 7–2. Fraction of length traversed versus time for linear frontal displacement showing the effect of mobility ratio. (*After Muskat, 1937.*)

From this it can be seen that the front either accelerates or decelerates according as $m > 1$ or $m < 1$. In the special case in which $m = 1$ the front moves with constant speed and also the pressure distribution is independent of x_f. This is seen in equations (7-20).

These results are presented graphically in Figure 7-2. These serve to emphasize the important role of mobility ratio in displacement problems.

7.20: Two-dimensional Displacement and Waterflood Sweep Patterns for Unit Mobility Ratio

Two-dimensional displacement problems are of special interest in the petroleum industry in connection with the waterflood technique of secondary recovery.

We consider a plane horizontal stratum of uniform thickness, permeability and porosity. This stratum is an oil reservoir which has been depleted by primary mechanisms, including natural water drive, so that it contains oil, water and gas at saturation S_o, S_w and S_g, respectively. In actuality these saturations would not be uniform throughout the reservoir. Gravity alone would lead to segregation which would be offset to some extent by capillary forces. In order to reduce the problem to be considered to a tractable form these saturations will be considered uniform throughout the reservoir initially.

Wells completed in this reservoir are located in some regular pattern in the plane of the stratum. Water is injected into selected wells at a constant rate. This injected water displaces the resident oil which is produced at the remaining wells. If gravity and capillary effects are ignored, the displacement can be assumed to take place in a piston-like manner. Thus the water-oil front is a surface of discontinuity in saturation having the form of a closed curve about each injection well in the plane of the stratum.

Even with these simplifications the problem is extremely complex. It is necessary to consider three regions in order to be even approximately realistic: the watered-out region, the region of the oil bank build up ahead of the injected water and the gas region ahead of the oil bank. Of course, each of these regions is a multifluid region. The water region, for example, would contain residual oil and gas.[10]

The simplest approximation to the real situation in waterflooding is achieved by considering the unflooded portion of the reservoir to contain mobile oil and immobile connate water, and the flooded region to contain mobile water and immobile residual oil. Then only two regions exist and only one front exists. This representation of the problem is amenable to mathematical solution.

The oil-water interface, or front, can be thought of as composed of fluid particles which always remain on the front no matter what the motion of the front may be. Thus, if the components of velocity of a large number of such particles, distributed more or less uniformly on the front, are known as functions of position and time, then the position of the front at any time can be obtained by integrating the set of equations

$$\left.\begin{aligned} \frac{dx_{p1}}{dt} &= v_{p1}(x_{p1},\, x_{p2},\, t) \\[2mm] \frac{dx_{p2}}{dt} &= v_{p2}(x_{p1},\, x_{p2},\, t) \end{aligned}\right\} \quad p = 1, 2, \cdots, M \qquad (7\text{-}26)$$

Here v_{p1} and v_{p2} are the velocity components of particle number p at the point x_{p1}, x_{p2} at time t.

From the discussion in section 7.10, we see that the vector velocity can be represented by

$$\hat{v}_p = \frac{\hat{v}_w}{\phi(1 - S_c - S_{ro})}$$

and $\hat{v}_w$, the volumetric flux density of water, is determined by the potential distribution according to Darcy's law.

The most general form of Darcy's law to be considered here is that corresponding to plane horizontal flow in a homogeneous anisotropic porous medium. The fluids are to be considered incompressible and gravity effects are assumed negligible. For these conditions

$$\hat{v}_w = -\frac{1}{\mu_w}\left(K_{w1}\frac{\partial p_w}{\partial x_1}\hat{i}_1 + K_{w2}\frac{\partial p_w}{\partial x_2}\hat{i}_2\right) \tag{7-27}$$

where the coordinate axes are along the principal axes of permeability. It is further assumed that relative permeability is independent of direction. Thus

$$k_w = \frac{K_{w1}}{K_1} = \frac{K_{w2}}{K_2} \tag{7-28}$$

Actually, no data on relative permeabilities of anisotropic porous media have been reported in the literature but this seems like a reasonable assumption and it does facilitate the mathematical treatment of the problem at hand.

It is also to be noted that in the watered-out region the relative permeability to water which applies is that corresponding to residual oil saturation. This is denoted by k_{wr}. Thus we have

$$\hat{v}_w = -k_{wr}\frac{K_1}{\mu_w}\left(\hat{i}_1\frac{\partial p_w}{\partial x_1} + \hat{i}_2\frac{K_2}{K_1}\frac{\partial p_w}{\partial x_2}\right) \tag{7-29}$$

Similarly we obtain

$$\hat{v}_o = -k_{oc}\frac{K_1}{\mu_w}\left(\hat{i}_1\frac{\partial p_o}{\partial x_1} + \hat{i}_2\frac{K_2}{K_1}\frac{\partial p_o}{\partial x_2}\right) \tag{7-30}$$

for the volumetric flux of oil in the unflooded region. Here k_{oc} is the relative permeability to oil at connate water saturation. Note that behind the front $p_w(x_1, x_2, t)$ applies while ahead of the front $p_o(x_1, x_2, t)$ applies, since only one fluid is mobile in each region.

Now we must have

$$v_{wn} = v_{on} \tag{7-31}$$

on the front. Hence, if capillary pressure is neglected, $p_o = p_w$ on the front and if we also assume

$$\frac{k_{wr}\mu_o}{k_{oc}\mu_w} = 1 \tag{7-32}$$

then the components of pressure gradient are continuous across the front. In this case one function, $p(x_1, x_2, t)$, describes the pressure distribution everywhere. (See section 7.21.) Combining Darcy's law, equation (7-29) or (7-30), and the equation of continuity for an incompressible fluid yields

$$\frac{\partial^2 p}{\partial x_1^2} + \frac{K_2}{K_1}\frac{\partial^2 p}{\partial x_2^2} = 0 \tag{7-33}$$

for this pressure function.

To complete the mathematical formulation of our simplified displacement problem, we write out our system of equations in terms of the dimensionless variables

$$\bar{x} = \frac{x_1}{L}, \quad \bar{y} = \frac{x_2}{L}\sqrt{\frac{K_1}{K_2}}, \quad \tau = \frac{qt}{\phi L^2 h\Delta S}\sqrt{\frac{K_1}{K_2}}, \quad \bar{p} = \frac{k_{oc}\sqrt{K_1 K_2}hp}{q\mu_o} \tag{7-34}$$

Here L is a characteristic linear dimension of the system (a well spacing, say), q is a characteristic flow rate, h is the vertical thickness of our horizontal stratum, and $\Delta S = 1 - S_c - S_{ro}$. (Cf. section 7.10.) In terms of these variables our equations are

$$\frac{\partial^2 \bar{p}}{\partial \bar{x}^2} + \frac{\partial^2 \bar{p}}{\partial \bar{y}^2} = 0 \tag{7-35}$$

and from equations (7-9), (7-26) and (7-29)

$$\left.\begin{array}{l} \dfrac{d\bar{x}_p}{d\tau} = -\dfrac{\partial \bar{p}}{\partial \bar{x}}(\bar{x}_p, \bar{y}_p, \tau) \\[3mm] \dfrac{d\bar{y}_p}{d\tau} = -\dfrac{\partial \bar{p}}{\partial \bar{y}}(\bar{x}_p, \bar{y}_p, \tau) \end{array}\right\} \quad p = 1, 2, \cdots, M \tag{7-36}$$

Here $\bar{x}_p$, $\bar{y}_p$, $p = 1, 2, \cdots m$ are the coordinates of M distinct points of the front.

Note that in view of equation (7-35) $\bar{p}$ depends on the time variable τ only through the time dependence of boundary conditions.

When considering a system of wells in a waterflood pattern we approximate the wells by point sources and sinks (line sources penetrating the stratum). Thus from section 4.50, we have in terms of the variables employed here

$$\bar{p} = \text{constant} - \frac{1}{4\pi}\ln\left[(\bar{x} - \bar{x}_i)^2 + (\bar{y} - \bar{y}_i)^2\right] \tag{7-37}$$

for a single-well injecting fluid at constant rate q at the point $\bar{x}_i$, $\bar{y}_i$, in a stratum of infinite areal extent. (This fixes q in equation 7-34.)

For a system of N wells located at points $\bar{x}_i$, $\bar{y}_i$, $(i = 1, 2, \cdots, N)$ with rates q_i we have by superposition

$$\bar{p}(\bar{x}, \bar{y}) = \text{constant} - \frac{1}{4\pi}\sum_{i=1}^{N}\frac{q_i}{q}\ln\left[(\bar{x} - \bar{x}_i)^2 + (\bar{y} - \bar{y}_i)^2\right] \tag{7-38}$$

for $\bar{p}$ at any point $\bar{x}$, $\bar{y}$. Note that for production wells the q_i are negative. Also note that any rates other than the artificially selected reference rate, q, may vary with time.

Here then for a system of wells in an infinite stratum we have completed the mathematical formulation. To determine the shape and position of the front about each injection well as a function of time we proceed as follows.

In the immediate neighborhood of an injection well the potential is essentially determined by equation (7-37) where $\bar{x}_i$, $\bar{y}_i$ are the coordinates of the well. Thus near the well the front is circular.* (In the $\bar{x}$, $\bar{y}$ coordinates only, in the x_1, x_2 system the front will be an ellipse unless $K_1 = K_2$). Hence we obtain from equations (7-36) and (7-37)

$$\frac{d\bar{r}_p}{d\tau} = \frac{1}{2\pi} \frac{1}{\bar{r}_p} \tag{7-39}$$

where

$$\bar{r}_p = [(\bar{x}_p - \bar{x}_i)^2 + (\bar{y}_p - \bar{y}_i)^2]^{1/2} \tag{7-40}$$

Integration then yields

$$\pi \bar{r}_p{}^2 = \tau_s \tag{7-41}$$

for constant rate. Here the well radius is neglected. For injection rate q_i this would be

$$\pi \bar{r}_p{}^2 = \frac{q_i}{q} \tau_s \tag{7-42}$$

Thus we begin by putting a small circle with radius given by equation (7-42) about each injection well. This, of course, fixes a starting value of τ also. On each of these circles (in the x, y plane), we place a number of equally spaced points. Then we approximate the equations of motion, equation (7-36) by

$$\left. \begin{array}{l} \bar{x}_p(\tau + \Delta\tau) = \bar{x}_p(\tau) - \Delta\tau \dfrac{\partial \bar{p}}{\partial \bar{x}} (\bar{x}_p, \bar{y}_p, \tau) \\[4mm] \bar{y}_p(\tau + \Delta\tau) = \bar{y}_p(\tau) - \Delta\tau \dfrac{\partial \bar{p}}{\partial \bar{y}} (\bar{x}_p, \bar{y}_p, \tau) \end{array} \right\} \tag{7-43}$$

or a higher order correct difference approximation. These equations can then be applied iteratively, for small $\Delta\tau$, to determine successive positions of the fronts about each injection well.

* The maximum radius for which this is true depends on the proximity of other wells.

For a regular pattern of wells such as the five-spot pattern shown in Figure 7-5 lines of symmetry connecting injection wells are streamlines. Thus, as in this pattern for example, flow is confined to a block for each production well. The area of this block is the pattern area, A. The production well produces fluid only from this block.

At any stage of the waterflood in such a system the watered-out, or flooded region, occupies an area A_w. At the instant that water reaches the production well, a certain area A_{wB} will have been flooded or swept. The areal sweep efficiency, $\mathcal{E}$, is defined as

$$\mathcal{E} = \frac{A_{wB}}{A} \tag{7-44}$$

This is usually multiplied by 100 and expressed as a percent of pattern area. The time at which water enters the production well is called the breakthrough time.

The sweep efficiency of a waterflood depends in the real situation on many factors: distributions of porosity and permeability, relative permeabilities and capillary pressure, geometry of the reservoir and well pattern, as well as injection and production rates. The formulation given above can show how sweep efficiency depends on some of these factors, in particular, geometry and flow rates of wells. Examples illustrating this are given in the following section. First, however, certain extensions of the mathematical formulation are required.

In section 4.33 it was pointed out that Laplace's equation, equation (7-35) in the present case, is invariant under a conformal transformation of coordinates. For flow geometries involving plane boundaries or plane discontinuities in permeability (in the $\bar{x}$, $\bar{y}$ system) the method of images can be used for solution of problems. But for curved boundaries conformal mapping must be used. A mapping must be used which transforms the problem to one with plane boundaries.

Thus suppose that

$$w = u + iv = w(\bar{z}) \tag{7-45}$$

where

$$\bar{z} = \bar{x} + i\bar{y} \tag{7-46}$$

is the required transformation. We have

$$\frac{\partial^2 \bar{p}}{\partial v^2} + \frac{\partial^2 \bar{p}}{\partial u^2} = 0 \tag{7-47}$$

and

$$u = u(\bar{x}, \bar{y})\Big\}$$
$$v = v(\bar{x}, \bar{y})\Big\} \tag{7-48}$$

In the w-plane, the solution is then

$$\bar{p} = \text{constant} - \frac{1}{4\pi} \sum_{i=1}^{N} \frac{q_i}{q} \ln [(u - u_i)^2 + (v - v_i)^2] \tag{7-49}$$

plus image terms, if required. Here

$$u_i = u(\bar{x}_i, \bar{y}_i)\Big\}$$
$$v_i = v_i(x_i, \bar{y}_i)\Big\} \quad i = 1, 2, \cdots, N \tag{7-50}$$

are the coordinates of the wells in the w-plane.

In the transformation of variables, we obtain as the equations of motion

$$\frac{d\bar{x}_p}{d\tau} = -\left[\frac{\partial \bar{p}}{\partial u} \frac{\partial u (\bar{x}_p, \bar{y}_p)}{\partial \bar{x}} + \frac{\partial \bar{p}}{\partial v} \frac{\partial u (\bar{x}_p, \bar{y}_p)}{\partial \bar{x}}\right]$$
$$\frac{d \bar{y}_p}{d\tau} = -\left[\frac{\partial \bar{p}}{\partial u} \frac{\partial u (\bar{x}_p, \bar{y}_p)}{\partial \bar{x}} + \frac{\partial \bar{p}}{\partial v} \frac{\partial u (\bar{x}_p, \bar{y}_p)}{\partial \bar{x}}\right] \tag{7-51}$$

for $p = 1, 2, \cdots, M$.

These equations are applied in the examples which follow.

The Isolated Two-Well Problem. This problem corresponds to a pilot or test flood. It consists of an injection well at $\bar{x} = \frac{1}{2}, \bar{y} = 0$ and a production well at $\bar{x} = -\frac{1}{2}, \bar{y} = 0$. The injection and production rates are equal. Here L is taken as the distance between the two wells and q is the injection rate.

Superposition of a point source and a point sink yields

$$\bar{p} = \text{constant} - \frac{1}{4\pi} \ln \frac{(\bar{x} - 0.5)^2 + \bar{y}^2}{(\bar{x} + 0.5)^2 + \bar{y}^2} \tag{7-52}$$

after rearrangement. This function employed in equations (7-43) is then used to compute the movement of points from the initial circle

$$(\bar{x} - 0.5)^2 + \bar{y}^2 = \epsilon^2 \tag{7-53}$$

where ϵ is the radius of the starting circle.

Streamlines are followed by each such point. The streamlines and front positions as computed in this manner are shown in Figure 7-3. The value of τ at breakthrough is 1.043.

Isolated Two-Well System in a Region Having a Discontinuity in Permeability. This problem is again a pilot flood problem and is chosen to illustrate the method of images discussed in section 4.51.

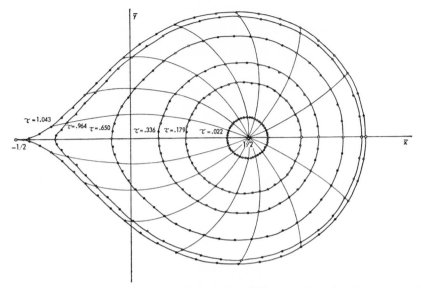

Figure 7–3. Front positions for various values of dimensionless time for an isolated two-well system.

We consider the $\bar{x}$, $\bar{y}$ plane bisected by a plane discontinuity passing through the origin at an angle of 45°. The equation of this line is $\bar{x} + \bar{y} = 0$.

It is assumed that for $\bar{x} + \bar{y} > 0$ the permeabilities are K_{1a} and K_{2a}, while for $\bar{x} + \bar{y} < 0$ the permeabilities are K_{1b} and K_{2b}. Also

$$\frac{K_{1a}}{K_{1b}} = \frac{K_{2a}}{K_{2b}} = \beta = \text{constant} \tag{7-54}$$

An injection well is located at $\bar{x} = +\frac{1}{2}$, $\bar{y} = 0$ and a production well at $\bar{x} = -\frac{1}{2}$, $\bar{y} = 0$. Image wells are required. The image of the injection well is located at $\bar{x} = 0$, $\bar{y} = -\frac{1}{2}$ and the image of the production well is at $\bar{x} = 0$, $\bar{y} = \frac{1}{2}$. The geometry of the system is shown in Figure 7-4.

By the method of images we have (for $\bar{x} + \bar{y} > 0$)

$$\bar{p} = \text{constant} - \frac{1}{4\pi}\left\{\ln\left[(\bar{x} - 0.5)^2 + \bar{y}^2\right] + \frac{\beta - 1}{\beta + 1}\ln\left[\bar{x}^2 + (\bar{y} + 0.5)^2\right]\right.$$
$$\left. - \frac{2}{\beta + 1}\ln\left[(\bar{x} + 0.5)^2 + \bar{y}^2\right]\right\} \tag{7-55}$$

and for $\bar{x} + \bar{y} < 0$

$$\bar{p} = \text{constant} - \frac{1}{4\pi}\left\{-\ln\left[(\bar{x} + 0.5)^2 + \bar{y}^2\right] - \frac{1 - \beta}{1 + \beta}\ln\left[\bar{x}^2 + (\bar{y} - 0.5)^2\right]\right.$$
$$\left. + \frac{2\beta}{1 + \beta}\ln\left[(\bar{x} - 0.5)^2 + \bar{y}^2\right]\right\} \tag{7-56}$$

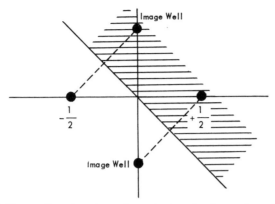

Figure 7–4. Two-well system in a plane stratum having a plane discontinuity in permeability, showing the image well locations.

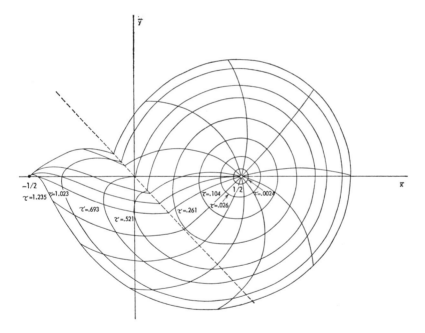

Figure 7–5. Front positions as a function of dimensionless time for the two-well system of Figure 7–4.

In this problem L is again the distance between wells and q is the constant injection rate. The injection and production rates are equal.

The small starting circle about the injection well is again used. For the case of $\beta = \frac{1}{10}$ the computed results are as shown in Figure 7-5. The breakthrough time is $\tau = 1.235$. Note that the injection well is in the region of lower permeability. Also note that the plane of discontinuity does not have the same slope in the x_1, x_2 plane unless the medium is isotropic.

Pattern Five-Spot. The well geometry for this problem consists of an infinite square array of wells arranged so that along any line there are alternate injection and production wells. The rates of injection and production are equal and the wells are equally spaced. This array, and the portion to be considered in the analysis are shown in Figure 7-6.

By symmetry the dotted-line boundaries of the shaded portion are streamlines and so represent planes across which no fluid flows. Treating this now as the plane of the complex variable $\bar{z} = \bar{x} + i\bar{y}$ the conformal transformation

$$w = u + iv = \sin \pi \bar{z} \tag{7-57}$$

transforms this strip into the upper half of the w-plane as shown in Figure 7-7.

Note that in this plane the u-axis is a streamline. In the original z-plane all wells on the boundary of the strip contributed one-half of their flow to the strip except the two corner wells. These two wells contributed only one-fourth of their flow to the strip. Thus, in the transformed plane all wells have the same rate except these two whose rates are one-half the rate of the other wells. Since the u-axis is a streamline the lower half-plane can

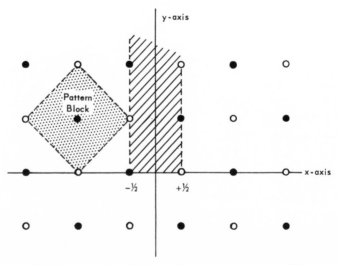

Figure 7–6. Infinite array of wells in the five-spot arrangement. The square region is a typical pattern element and the vertical strip is the region employed in the mathematical solution.

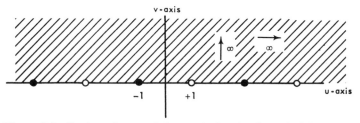

Figure 7-7. Conformal map of the vertical strip shown in Figure 7-6.

be supplied as the image of the upper half-plane and we have an infinite line of wells in the infinite plane.

The transformation, equation (7-57), gives

$$\left.\begin{array}{l} u = \sin \pi \bar{x} \cosh \pi \bar{y} \\ v = \cos \pi \bar{x} \sinh \pi \bar{y} \end{array}\right\} \tag{7-58}$$

Hence the well coordinates in the w-plane are

$$v_k = 0$$

$$u_k = \begin{cases} \cosh \pi (k - 1) & k = 1, 2, \cdots . \\ -\cosh \pi (k + 1) & k = -1, -2, \cdots . \end{cases} \tag{7-59}$$

By superposition we have then

$$\begin{aligned} \bar{p} = \text{constant} & - \frac{1}{4\pi} \left\{ \sum_{k=2}^{\infty} (-1)^{k+1} \ln [(u - u_k)^2 + v^2] + \tfrac{1}{2}\ln [(u - u_1)^2 + v^2] \right\} \\ & + \frac{1}{4\pi} \left\{ \sum_{k=-2}^{\infty} (-1)^{-k+1} \ln [(u - u_k)^2 + v^2] + \tfrac{1}{2}\ln [(u - u_1)^2 + v^2] \right\} \end{aligned} \tag{7-60}$$

for the pressure function. This, together with equation (7-58) and (7-59) above are to be used in equation (7-51) to compute the front position at various values of τ.

Since a high degree of symmetry exists here only a portion of one front needs to be computed. Thus an initial quarter circle about the injection well at $\bar{x} = \tfrac{1}{2}$, $\bar{y} = 0$ is used. It is also noted that wells far removed from the vicinity of this front have negligible effect on its motion. Thus retaining only the first ten terms in the series in equation (7-60) yields values of $\bar{p}$ correct to seven significant figures for the region of the front.

The sweep pattern computed by this procedure is shown in Figure 7-8. The breakthrough time is $\tau = 1.431$. Note that here q is the injection rate per injection well and L is the distance from injection well to production well. The areal sweep efficiency at breakthrough is 71.55%.

Discussion of Examples. The examples given above illustrate a very

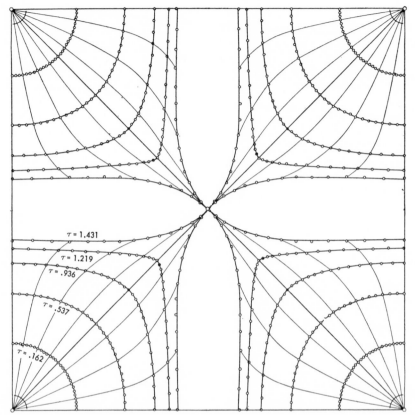

Figure 7–8. Front positions as a function of dimensionless time for the five-spot pattern.

versatile method for treating a restricted class of waterflood problems. It must be noted that when treating an anisotropic system by the above method the coordinate axes must be selected to coincide with the principal axes of permeability.

This method can be extended to cover a great variety of problems of practical interest, but the limitations are severe. The restriction to unit mobility ratio is very critical and is discussed further in the following section.

With proper restrictions the above method can be extended to compressible liquids by employing the point-source solution discussed in section 5.30, in particular, equation (5-49). The method can also be extended to three-dimensional systems. However, conformal mapping cannot be used in these extensions.

One important application of the methods described here is to the evaluation of pilot waterfloods. Thus for a known well geometry and known injection and production rates one can compute flood patterns and corresponding breakthrough times for various assumed orientations of principal axes of permeability, various values of K_1/K_2 and various values of $\phi L^2 h \Delta s$. When a set of parameters is found which produces breakthrough times matching those actually observed, these can be used to predict the oil recovery for a full-scale flood.

7.21: The Effect of Mobility-Ratio on Areal Sweep Efficiency

The treatment of frontal displacement in a homogeneous medium is extremely complex if the mobility ratio, $k_{wr}\mu_o/k_{oc}\mu_w$ is not unity. In the water region we have*

$$\frac{\partial^2 p_w}{\partial x_1{}^2} + \frac{K_2}{K_1}\frac{\partial^2 p_w}{\partial x_2{}^2} = 0 \tag{7-61}$$

as before, and in the unflooded oil region

$$\frac{\partial^2 p_o}{\partial x_1{}^2} + \frac{K_2}{K_1}\frac{\partial^2 p_o}{\partial x_2{}^2} = 0 \tag{7-62}$$

Now the boundary conditions on the front are taken as

$$\left.\begin{array}{c} p_o = p_w \\[2mm] v_{on} = v_{wn} \end{array}\right\} \quad \text{on front} \tag{7-63}$$

Here capillary and gravity effects are still ignored. The second boundary condition above, continuity of volume flux normal to the boundary, is the condition derived in section 7.10. Note that the fluids are again treated as incompressible.

Before proceeding further with methods of treating such problems, we will express our equations in terms of the same dimensionless variables used in the previous section. Thus, let

$$\bar{x} = \frac{x}{L}, \quad \bar{y} = \frac{y}{L}\sqrt{\frac{K_1}{K_2}}, \quad \tau = \frac{qt}{\phi L^2 h \Delta s}\sqrt{\frac{K_1}{K_2}} \tag{7-64}$$

and define

$$\bar{p} = \frac{k_{oc}\sqrt{K_1 K_2}h p_o}{q\mu_o}, \quad \bar{p}_w = \frac{k_{oc}\sqrt{K_1 K_2}h p_w}{q\mu_o} \tag{7-65}$$

Then our differential equations are

* Relative permeability is again assumed isotropic. (Cf. equation 7-28.)

$$\frac{\partial^2 \bar{p}_w}{\partial \bar{x}^2} + \frac{\partial^2 \bar{p}_w}{\partial \bar{y}^2} = 0 \qquad (7\text{-}66)$$

in the water region, and

$$\frac{\partial^2 \bar{p}_o}{\partial \bar{x}^2} + \frac{\partial^2 \bar{p}_o}{\partial \bar{y}^2} = 0 \qquad (7\text{-}67)$$

in the oil region. The first-boundary condition on the front is

$$\bar{p}_o = \bar{p}_w \qquad \text{(on the front)} \qquad (7\text{-}68)$$

The form of the second-boundary condition on the front, continuity of normal flux, is deduced as follows.

Let

$$\widehat{ds} = \hat{\mathbf{1}}_x dx_1 + \hat{\mathbf{1}}_y \, dx_2 \qquad (7\text{-}69)$$

be a directed line segment along the front, $\hat{\mathbf{1}}_x$ and $\hat{\mathbf{1}}_y$ being unit vectors parallel to the respective axes. Then the volume flux of water across this line segment is

$$v_{wn} \, ds = -\frac{k_{wr}}{\mu_w} \left(K_1 \frac{\partial p_w}{\partial \bar{x}_1} dx_2 - K_2 \frac{\partial p_w}{\partial x_2} dx_1 \right) \qquad (7\text{-}70)$$

Then substituting our dimensionless variables yields

$$v_{wn} \, ds = -\frac{q}{h} \frac{k_{wr}\mu_o}{k_{oc}\mu_w} \left(\frac{\partial \bar{p}_w}{\partial \bar{x}} d\bar{y} - \frac{\partial \bar{p}_w}{\partial \bar{y}} d\bar{x} \right) \qquad (7\text{-}71)$$

Similarly

$$v_{on} \, ds = -\frac{g}{h} \left(\frac{\partial \bar{p}_o}{\partial \bar{x}} d\bar{y} - \frac{\partial \bar{p}_w}{\partial \bar{y}} d\bar{x} \right) \qquad (7\text{-}72)$$

Thus the second-boundary condition on the front is

$$\frac{k_{wr}\mu_o}{k_{oc}\mu_w} \frac{\partial \bar{p}_w}{\partial \bar{n}} = \frac{\partial \bar{p}_o}{\partial \bar{n}} \qquad \text{(on the front)} \qquad (7\text{-}73)$$

where $\bar{n}$ is distance measured normal to the boundary in the $\bar{x}$, $\bar{y}$ coordinate system.

Note that for unit mobility ratio this, in conjunction with $\bar{p}_o = \bar{p}_w$, shows that one function $\bar{p}$ suffices. This is the case treated in the last section. Also note that for $K_1 = K_2$, that is an isotropic medium, this boundary condition applies in the x_1, x_2 coordinate system also, but only for such media is this true.

The system of equations: (7-66), (7-67), (7-68) and (7-73), coupled with boundary conditions at the wells and on all fixed boundaries governs the distribution of pressure. Since one boundary, the front, is moving the problem cannot be solved without simultaneously determining the evolution of the front with time. This dependence on the shape and position of the front makes the pressure-distribution time dependent even for constant boundary conditions and incompressible fluids. This was demonstrated in the linear case given in section 7.11.

In principle, one could solve the pressure problem simultaneously with equation (7-11) for F defining the position of the front. In the present case, this could be written as

$$\frac{\partial \bar{p}_w}{\partial \bar{x}} \frac{\partial F}{\partial \bar{x}} + \frac{\partial \bar{p}_w}{\partial \bar{y}} \frac{\partial F}{\partial \bar{y}} = \frac{\partial F}{\partial \tau} \tag{7-74}$$

Actually this can be carried out by methods of finite differences on a large high-speed computer, but the procedure is very complex.

Since this entire formulation is only an approximation to the real physical situation which must involve relative permeabilities and capillary pressures, and no clear-cut front, the best approach for any elaborate computation is that of two-dimensional, two-phase flow as carried out by Douglas, Peaceman and Rachford.[6] This was briefly described in section 6.50. The reader should consult the reference for more details.

In spite of its limitations the frontal displacement problem as formulated here is of practical interest. Approximate solutions of the problem have been obtained with electrical analog models by Aronofsky.[1] Aronofsky has also solved some examples by numerical techniques. Other workers[4, 9, 12] have obtained solutions with models. The goal of all these investigators is to determine the dependence of areal-sweep efficiency on mobility ratio for selected well patterns. Some typical results as obtained by Aronofsky for a line-drive array of wells are shown in Figure 7-9. The well pattern is indicated in the figure.

Generally speaking the areal-sweep efficiency always decreases as the mobility ratio increases. That is, if the mobility of the displacing fluid is made less than the mobility of the resident fluid.

7.30: Filtration and the Deposition of Solids

An important class of moving-boundary problems arises from the filtration of a fluid suspension of solid particles by porous media. For example, in drilling of oil wells the bore hole is filled with drilling mud. This mud is essentially a suspension of clay particles in water. At some depth, the hydrostatic pressure in the bore hole is greater than the pressure in a fluid filling a stratum of porous rock penetrated by the hole. Thus, the water filters into the porous rock depositing a cake of clay particles on the wall

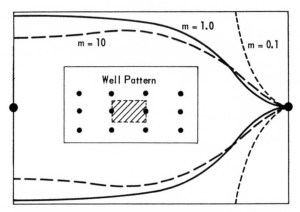

Figure 7–9. Front positions at water breakthrough for the direct-line drive show-ing the effect of mobility ratio. (*After Aronofsky, 1952.*)

of the bore hole. A similar process occurs in squeeze cementing operations[2] designed to plug undesired completion perforations in the well casing.

Filter cakes formed in this manner on a porous surface are porous and also permeable. However, their porosity and permeability will, in general, depend on the pressure gradient to which they are exposed since they are usually very compressible.

If the particle size of the suspension is less than the average pore size of the porous surface on which the filter cake is formed, then some solid particles will penetrate into the porous surface.

A mathematical description of such deposition processes can be formulated as follows.

The total volume of suspension, which is assumed homogeneous, is composed of the volume of solid particles, V_s and the volume of liquid, V_l. Thus, the fraction by volume of solids is

$$f_s = \frac{V_s}{V_s + V_l} \tag{7-75}$$

During the filtration and deposition process V_s is decreased by an amount, dV_s, during a time, dt. During the same period V_l is decreased by dV_l. If we assume that the composition of the suspension remains unchanged, then df_s is zero and we have

$$dV_s = \frac{f_s}{1 - f_s} dV_l \tag{7-76}$$

In the filtration process solid particles are being deposited within the pores of the filtering medium, or in the form of a filter cake on the outer surface of the filtering medium. Let δA be an element of area on which particles are deposited, and dx_c be the increase in thickness of the deposition region occurring during time dt. If this region is within the filtering medium, then only a fraction ϕ of the volume $\delta A\,dx_c$ is available to the particles. Here ϕ is the porosity of the filtering medium. Within the available volume the solid particles are deposited with a porosity, ϕ_c. Then the total volume of solids deposited on δA during dt is

$$dV_s = \phi(1 - \phi_c)\delta A\,dx_e \tag{7-77}$$

During the time dt a volume of liquid flows through the area δA. If the volume flow rate is v_n per unit area normal to δA, we have

$$dV_l = -v_n\delta A\,dt \tag{7-78}$$

The minus sign is required because v_n is directed opposite to the outward drawn normal to δA.

Combining equations (7-77) or (7-78) and (7-76) yields

$$\frac{dx_c}{dt} = - \frac{f_s}{\phi(1 - \phi_c)(1 - f_s)}\,v_n \tag{7-79}$$

as the law of deposition. (Note that this is not exactly the same as in reference 2. The form given in the cited paper is in error.) Here we put $\phi = 1$ for filter-cake formation outside the filtering medium.

The factors

$$\omega_i = \frac{f_s}{\phi(1 - \phi_c)(1 - f_s)} \tag{7-80}$$

and

$$\omega_e = \frac{f_s}{(1 - \phi_c)(1 - f_s)} \tag{7-81}$$

are the internal and external deposition factors, respectively. Since the filter cake is always at least slightly compressible ϕ_c varies with pressure differential and hence the deposition factor is not exactly constant. However, treating ω_i and ω_e as constants does not lead to significant errors for deposition under constant pressure differential.

The flow of fluid within the deposition region obeys Darcy's law with a permeability K_c for external filter cake. In the internal deposition region the permeability is less. Approximately, the permeability is ϕK_c for the internal case. K_c also depends on pressure differential because of the compressibility of the filter cake.

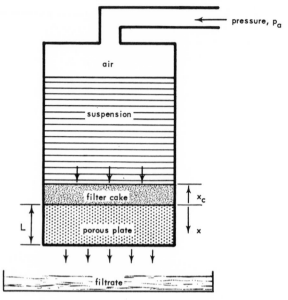

Figure 7–10. Filtration apparatus.

To show how certain of the factors defined above can be measured and to illustrate the macroscopic features of such filtration processes consider the case of linear filtration.

We consider a chamber containing a suspension in contact with a porous permeabile plate. The plate is in a horixontal plane as indicated in Figure 7-10.

Only external deposition is considered. Both the filtrate and filter cake are considered incompressible. We have by Darcy's law

$$q = -\frac{K_c A}{\mu}\left(\frac{\partial p}{\partial x} - \rho g\right) \tag{7-82}$$

where A is the area of the plate, ρ and μ are the density and viscosity of the filtrate, and q is the volumetric flow rate of filtrate in the plus x direction. (Rate is downward as indicated in the figure. A similar expression with K_c replaced by K, the permeability of the plate, applies within the plate.)

Since the fluid is incompressible

$$\frac{\partial q}{\partial x} = 0 \tag{7-83}$$

is the equation of continuity. Thus

$$\frac{\partial^2 p}{\partial x^2} = 0 \tag{7-84}$$

everywhere. The solution of this is

$$\left. \begin{array}{ll} p = ax + b & \text{for} \quad -x_c < x < 0 \\ p' = a'x + b' & \text{for} \quad 0 < x < L \end{array} \right\} \tag{7-85}$$

The boundary conditions are

$$\left. \begin{array}{l} p(-x_c) = p_a + \rho_m gh \\ p'(0) = p(0) \\ p'(L) = 0 \\ -K_c \left(\dfrac{\partial p}{\partial x} - \rho g \right)_{x=0} = -K \left(\dfrac{\partial p'}{\partial x} - \rho g \right)_{x=0} \end{array} \right\} \tag{7-86}$$

Here p_a is the air pressure above the suspension, h is the depth of the suspension (surface of suspension to surface of filter cake) and ρ_m is the density of the suspension. When these conditions are used to evaluate the constants a, b, a' and b' there results

$$p = -\frac{p_a + \rho_m gh + \left(1 - \dfrac{K_c}{K} \right) \rho g L}{x_c + \dfrac{K_c}{K} L} x + b \tag{7-87}$$

and then by equation (7-82)

$$q = \frac{K_c A}{\mu} \left[\frac{p_a + \rho_m gh + \rho g\, (L + x_c)}{x_c + \dfrac{K_c}{K} L} \right] \tag{7-88}$$

Then according to equation (7-79)

$$\frac{dx_c}{dt} = \frac{K_c}{\mu} \left[\frac{p_a + \rho_m gh + \rho g(L + x_c)}{x_c + \dfrac{K_c}{K} L} \right] \frac{f_s}{(1 - \phi_c)(1 - f_s)} \tag{7-89}$$

since the position of the cake surface was here taken as $-x_c$.

From the definition of h and f_s, we can write

$$h = h_i - x_c - \frac{Q}{A(1 - f_s)} \tag{7-90}$$

where

$$Q = \int_0^t q \, dt \tag{7-91}$$

is the cumulative filtrate formed and h_i is the initial value of h. But, from equation (7-79) we have by integration

$$Q = \frac{A(1 - f_s)(1 - \phi_c)}{f_s} \, x_c \tag{7-92}$$

for $Q = x_c = 0$ at $t = 0$.

Equation (7-90) with Q given by (7-92) can be substituted for h in equation (7-89). Then equation (7-89) can be integrated for $p_a = $ constant. However, in the usual case p_a is quite large compared to the gravity terms. Hence for large p_a

$$\frac{dx_c}{dt} \approx \frac{K_c p_a \omega_e}{\mu \left(x_c + \dfrac{K_c}{K} L \right)} \tag{7-93}$$

Then integration yields for $x_c = 0$ at $t = 0$

$$x_c = -\frac{K_c}{K} L + \left[\left(\frac{K_c}{K} L \right)^2 + \frac{2K_c p_a \omega_e}{\mu} t \right]^{1/2} \tag{7-94}$$

Now if Q and x_c are measured at some time t then from equation (7-92) ω_e can be computed as

$$\omega_e = \frac{Q}{A x_c} \tag{7-95}$$

Thus this number along with the values of L, p_a, K, μ, x_c and t can be employed with equation (7-94) to compute the value of K_c, i.e.

$$K_c = \frac{x_c^2 \mu K}{2 (p_a K \omega_e t - \mu x_c L)} \tag{7-96}$$

Note that Q includes not only the filtrate expelled from the filter plate but also that in the filter plate. This correction is eliminated by starting with the plate saturated with liquid.*

It is to be noted that for small values of L, x_c and Q both increase as the square root of the time. This time dependence has often been noted in the literature for large values of t.

Some typical data obtained in this manner are shown in Figures 7-11

---

* Note that the value of Q measured will still be less than the actual filtrate formed by the amount in the filter cake itself, i.e. $\phi_c A x_c = \Delta Q$.

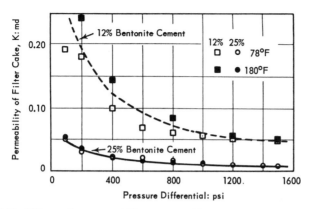

Figure 7–11. Filter-cake permeability versus filtration pressure differential for two bentonite clay-cement slurries. (*After Binkley et al., 1958.*)

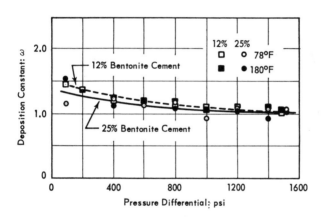

Figure 7–12. Deposition factor versus filtration pressure differential for two bentonite clay-cement slurries. (*After Binkley et al. 1958.*)

and 7-12. These show the dependence of K_c and ω_e on pressure differential[2] for two cement suspensions. One suspension is of cement and 12 % bentonite clay in water, the other contains 25 % clay. It is to be noted that although K_c varies greatly with pressure differential, ω_e is nearly constant.

In a general multidimensional deposition problem the boundary condition to be employed at the interface of filter cake and suspension is

$$\frac{ds}{dt} = -\omega_e \hat{v} \cdot \hat{n} \tag{7-97}$$

Here ds/dt is the rate of growth of the interface normal to itself, $\hat{v}$

is the volume flux density of filtrate at the interface and $\hat{n}$ is a unit vector parallel to the outward normal to the interface. Thus filter-cake growth on any shape of filter surface can be studied.[2]

7.40: Frontal Instability and Viscous Fingering

In the previous sections of this chapter various aspects of frontal displacement have been considered. Several important effects of mobility ratio were discussed in connection with waterflooding of petroleum reservoirs. However, one very important phenomenon associated with frontal movement, which is strongly dependent on mobility ratio, has been neglected. This phenomenon is frontal instability and the associated formation of what are usually called viscous fingers.

Instability of a displacement front can most easily be understood by considering a somewhat oversimplified model. Consider a linear frontal displacement in which the displacing fluid is more mobile than that being displaced. From our previous analysis the front should remain a plane surface throughout the displacement. But suppose that a tiny region of the porous medium is not homogeneous. If this tiny region is more permeable than the surrounding region, then as the front approaches this region that part nearest to the tiny region will move more rapidly. This gives rise to a small "bump" on the otherwise plane front

To discover the subsequent history of this bump consider a tube parallel to the direction of flow containing the bump, as indicated in Figure 7-13. This tube is now treated as an isolated linear system while the other portion of the system constitutes another isolated linear system. Denote the position of the front relative to the inflow end in this second undisturbed region by x_f and that of the front in the disturbed system by $x_f + \epsilon$. Here ϵ is the length of the bump, which is considered infinitesimal.

From the discussion of the linear displacement in section 7.11 we have

$$\frac{dx_c}{dt} = \frac{K_{wr}\Delta p}{\mu_w \phi (1 - S_c - S_{ro})[mL + (1 - m)x_f]} \tag{7-98}$$

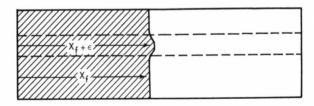

Figure 7–13. Simplified model of frontal instability.

and

$$\frac{d(x_f + \epsilon)}{dt} = \frac{K_{wr}\Delta p}{\mu_w \phi (1 - S_c - S_{ro})[mL + (1 - m)(x_f + \epsilon)]} \tag{7-99}$$

where

$$m = \frac{K_{wr}\mu_o}{K_{oc}\mu_w} \tag{7-100}$$

is the mobility ratio. From these, there results

$$\frac{d\epsilon}{dt} = \frac{-K_{wr}\Delta p(1 - m)\epsilon}{\mu_w \phi (1 - S_c - S_{ro})[mL + (1 - m)x_f]^2} \tag{7-101}$$

provided $\epsilon \ll x_f$.

Here then we see that ϵ grows exponentially with time after inception of the bump if $m > 1$, or decays exponentially with time if $m < 1$. For $m = 1$ ϵ does not change with time.

This simple analysis shows that, if the displacing fluid is more mobile than the displaced fluid, any small perturbation of the front gives rise to irregularities which grow very rapidly. Such irregularities usually take the form of "fingers" extending from the front and hence the name viscous fingering. If $m < 1$ the front is stable while for $m > 1$ the front is unstable.

This simple treatment neglects the effects of gravity and capillarity, both of which usually tend to minimize viscous fingering. In particular, since in oil-water systems the oil is less dense than the water, but is usually less mobile, a waterflood moving up-dip in an oil reservoir will have gravity effects tending to reduce fingering. In a down structure flood gravity would accentuate fingering.

A critical point is the fact that some type of inhomogeneity or perturbation is required to initiate the instability and hence fingering. Here the microscopic nature of porous media must be considered. Natural porous media have a microscopically random porous structure. The mathematical description employed above is macroscopic. Thus an infinity of random perturbations is present in the most uniform porous medium imaginable. Consequently, fingering will always occur if $m > 1$. However, gravity and capillarity may act to eliminate fingers as they are formed. Furthermore, in the case of miscible fluids treated in another chapter, diffusion may also serve to reduce fingers.

The process by which gravity serves to eliminate fingering is as follows. For a waterflood directed up-structure a finger when formed has acting to oppose its growth an additional hydrostatic head, essentially $\Delta \rho g \epsilon$. For a more precise evaluation of this effect, the treatment of the linear displacement of section 7.11 is repeated for a linear system tilted at an angle θ above the horizontal. When gravity is included we have the boundary condition

$$p_o = p_w$$

$$\frac{K_{wr}}{\mu_w}\left[\frac{\partial p_w}{\partial x} + \rho_w g \, \sin \theta\right] = \frac{K_{oc}}{\mu_o}\left[\frac{\partial p_o}{\partial x} + \rho_o g \, \sin \theta\right] \tag{7-102}$$

at $x = x_f$. Otherwise the problem is as before. Note that x is measured along the tube and variations across the tube are ignored.

In this case there results

$$\frac{dx_f}{dt} = \frac{K_{wr}[\Delta p + m\Delta\rho g(L - x_f) \sin \theta]}{\mu_w\phi(1 - S_c - S_{ro})[mL + (1 - m)x_f]} \tag{7-103}$$

for the rate of advance of the front.

Repeating our derivation for the growth of a finger of length ϵ yields

$$\frac{d\epsilon}{dt} = - \frac{K_{wr}\epsilon}{\mu_w\phi(1 - S_c - S_{ro})} \frac{(1 - m)\Delta p + m\Delta\rho gL \sin \theta}{[mL + (1 - m)x_f]^2} \tag{7-104}$$

For zero growth rate we put $d\epsilon/dt = 0$, let $\phi(1 - S_c - S_{ro})dx_f/dt = v_w$ denote the flow rate per unit area for water and use equation (7-103) to eliminate Δp in equation (7-104). There results then

$$v_w \leq \frac{m}{m - 1} \frac{K_{wr}\Delta\rho g \sin \theta}{\mu_w} \tag{7-105}$$

as the condition for no fingers to form. A more detailed theory of this critical velocity is given by Hawthorne.[7]

Observe that this indicates the flood should be directed down-structure if the density of the displacing fluid is less than that of the resident fluid and $m > 1$. Also observe that for $m = 1$ this predicts no fingering at any fluid velocity.

The existence of the critical maximum velocity for no finger formation as predicted above is confirmed by laboratory experiments. Blackwell, Rayne and Terry[3] have reported results confirming this result. Figure 7-14 shows a plot of recovery at breakthrough for miscible fluids versus flow rate. For these data the mobility ratio was $m = 5/1$, the density difference $\Delta\rho = -0.104 \, \text{gm/cm}^3$ and the system was vertical. The flood was directed downward.

For a linear system the recovery at breakthrough should be 100 % if no fingering occurs. If fingering occurs the recovery will be less than 100 % at breakthrough.

To illustrate the serious extent to which fingering may occur for high

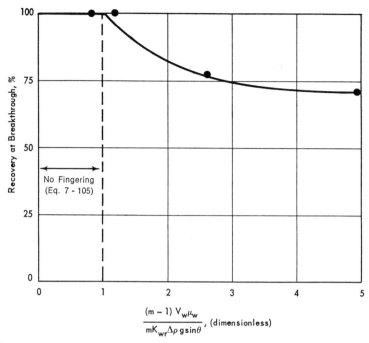

Figure 7-14. Recovery versus displacement velocity factor showing the existence of critical displacement velocity. (*After Blackwell et al., 1958.*)

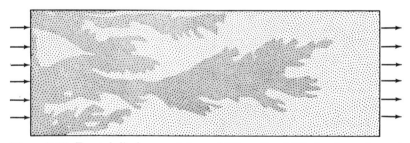

Figure 7-15. Frontal displacement for mobility ratio of 383 showing extreme instability and severe viscous fingering. (*After Blackwell et al., 1958.*)

mobility ratios consider Figure 7-15. This shows the form of the flood front in an unconsolidated sand packed as uniformly as possible. The view is perpendicular to the thickness of the sand pack. In this case the mobility ratio was 383 to 1 and the fluids were miscible (zero capillary pressure).

At the present time the problem of viscous fingering is one of the most prominent problems in the petroleum industry. Because of the random nature of natural porous structures it is certain that any adequate theory must be of a type which combines a macroscopic description of the flow with a statistical description of the porous medium. Scheidegger[11] has recently published a first attempt at such a theory. In its published form Scheidegger's theory is limited by mathematical approximations introduced to achieve linearity in the equations, but the basic approach seems valid. Chuoke[5] has treated the same problem in a somewhat different manner.

Further considerations of the fingering and instability problem are given in the next chapters in connection with the simultaneous flow of miscible fluids and the theory of models.

EXERCISES

1. Solve the problem of radial frontal displacement of oil by water between the concentric circular boundaries r_1 and r_2. Use $P = P_1$ on r_1 and $P = P_2$ on r_2.

2. Show that for a frontal displacement process with either $m = 0$ or $m = \infty$ the displacement front is a surface of constant pressure. Also either the region ahead of the front or the region behind the front is a region of uniform pressure, depending on whether $m = 0$ or $m = \infty$.

3. In the formulation of frontal displacement for $m = 1$ represented by equation (7-35) and (7-36) with variables defined by equation (7-34) show that for the isolated two well problem the dimensionless breakthrough time, τ, is just the fraction of the circular area πL^2 flooded at breakthrough when $K_1 = K_2$. Generalize this result for other systems of wells and for $K_1 \neq K_2$.

References

1. Aronofsky, J. S., *Trans. AIME*, **195**, 15 (1952).
2. Binkley, G. W., Dumbauld, G. K., and Collins, R. E., *Trans. AIME*, **213**, 51 (1958).
3. Blackwell, R. J., Rayne, J. R., and Terry, W. M., *Trans. AIME*, **216**, 1 (1959).
4. Burton, M. B., Jr., Crawford, P. B., *Trans. AIME*, **207**, 333 (1956).
5. Chuoke, R. L., *Trans. AIME*, **216**, 188 (1959).
6. Douglas, J. Jr., Peaceman, D. W., and Rachford, H. H., Jr., *Trans. AIME*, **216**, 297 (1959).
7. Hawthorne, R. G., *J. Petroleum Technol.* 82: April 1960.
8. Muskat, M., "Flow of Homogeneous Fluids through Porous Media," McGraw-Hill Book Co., New York (1937); J. W. Edwards, Inc., Ann Arbor (1946).
9. Ramey, H. J., and Nabor, G. W., *Trans. AIME*, **201**, 35 (1954).
10. Rapoport, L. A., Carpenter, C. W., Jr., and Leas, W. J., *Trans. AIME*, **213**, 113 (1958).
11. Scheidegger, A. E., *Phy. of Fluids*, **3**, 94 (1960).
12. Slobod, R. L., and Caudle, B. H., *Trans. AIME*, **195**, 265 (1952).

8. SIMULTANEOUS LAMINAR FLOW OF MISCIBLE FLUIDS

8.10: Miscible and Immiscible Fluids, Fick's Law of Diffusion

In all considerations of multifluid systems of the previous chapters the fluids were considered as immiscible. That is the fluids would not mix in a physical sense. If the interfacial tension between two fluids is non-zero the fluids do not mix; a distinct fluid-fluid interface always separates the fluids. If the interfacial tension between two fluids is zero then a distinct fluid-fluid interface does not exist and the fluids are miscible.

If two fluids are miscible then molecules of one fluid can diffuse into the other fluid. This is a spontaneous process. It can be thought of as occurring by the following mechanism.

Consider two fluids brought into contact at a plane. Within either fluid the molecules have a random motion which is dependent upon the absolute temperature. This motion is isotropic; that is, in any homogeneous region there are equal numbers of molecules moving in all directions with the same distribution of velocity.

At the plane of separation there are molecules of kind 1 on the left, say, and molecules of kind 2 on the right. Due to the random motion some molecules of kind 1 cross the plane to the right and some of kind 2 cross to the left. This process expands in both directions until a homogeneous mixture of the two kinds of molecules exists. This process is termed "molecular diffusion."

If the fluids were immiscible, the molecules of kind 1 attempting to move to the right across the plane of separation would be acted on by a force field in the neighborhood of the interface which would restrain them. Thus no mixing by diffusion would occur.

The heuristic description of diffusion given leads to the law of diffusion termed Fick's law.[7] It is evident that the rate of movement of molecules should depend on the relative concentration. Thus the rate of movement across a plane should depend on the difference in concentration across the plane. More specifically, the rate of movement can be represented by

$$\frac{dn}{dt} = -D'A \frac{\partial C'}{\partial x} \tag{8-1}$$

Here dn/dt is the number of molecules crossing the area A per unit time in the direction of increasing x, C' is the concentration in molecules per volume of molecules of the kind being considered and D' is a factor called the diffusion coefficient, or the diffusion constant. The dimensions of D' are length squared per unit time (cm^2/sec in c.g.s. units).

Generally D' is not exactly a constant. Not only does D' depend upon the absolute temperature but it also varies somewhat with concentration. Furthermore the value of D' for a particular kind of molecule depends upon what other kinds of molecules are present. However, in the majority of applications D' can be treated as a constant for a particular problem.

Often the diffusion of a material substance is expressed in mass per unit time. If equation (8-1) is multiplied by M/L where M is the molecular weight of the diffusing substance and L is Avagadro's number, the number of molecules per mole, there results

$$\frac{dm}{dt} = -\frac{MD'}{L} A \frac{\partial C'}{\partial x} \tag{8-2}$$

Here dm/dt is mass per unit time diffusing across A. If the concentration, C', is expressed in the more common units of mass of diffusing material per mass of total substance, we have

$$C = \frac{MC'}{L\rho} \tag{8-3}$$

where ρ is the mass density of total substance. If ρ is treated as being constant, independent of composition of the substance, then Fick's law can be written in the form

$$\frac{dm}{dt} = -DA \frac{\partial C}{\partial x} \tag{8-4}$$

where $D = \rho D'$ is the diffusion constant expressed in mass/length-time. The common units are gm/cm-sec.

Fick's law can be written for a general geometry in the multidimensional case, as

$$\hat{m} = -D\nabla C \tag{8-5}$$

This is the form most often employed in applications to flow problems. Here $\hat{m}$ is the mass flux density vector, mass per unit time per unit area.

8.20: Miscible Displacement in a Capillary Tube

To illustrate some of the microscopic features of miscible displacement in porous materials we consider first the problem of the displacement of a fluid from a straight circular capillary tube by another fluid which is miscible with the resident fluid.

If the two fluids have the same viscosity and density the distribution of fluid velocity within the tube does not depend on the distribution of the two fluids within the tube. For slow steady flow at the mean velocity, $\bar{v}$, the velocity at a point a distance r from the axis of the tube is[12]

$$v(r) = 2\bar{v} \left(1 - \frac{r^2}{a^2}\right) \qquad (8\text{-}6)$$

where a is the radius of the tube. The fluid at the wall of the tube does not move and the fluid on the axis of the tube has the maximum speed. Thus, if a group of marked particles lies on a plane perpendicular to the axis at time zero they will lie on the surface of a paraboloid of revolution at any later time; this by convection alone.

If at time $t = 0$ the concentration distribution of injected fluid is $C(x, r)$, where x is measured along the axis, then at time t the concentration is

$$C = C(x - vt, r) \qquad (8\text{-}7)$$

where v is given by equation (8-6); again by convection alone. Thus, convection alone produces a dispersion of injected fluid.

Since the fluids are miscible a dispersion of the injected fluid also occurs by diffusion. The equation governing this diffusion is deduced by requiring the conservation of mass of injected fluid.

Consider the annular tube within the fluid of length Δx between x and $x + \Delta x$, inner radius r and outer radius $r + \Delta r$. Equating the net mass flow into the tube, both by diffusion and convection, to the rate of increase of mass content of injected fluid, there results

$$D\left(\frac{\partial^2 C}{\partial r^2} + \frac{1}{r}\frac{\partial C}{\partial r} + \frac{\partial^2 C}{\partial x^2}\right) = \rho\frac{\partial C}{\partial t} + 2\rho\bar{v}\left(1 - \frac{r^2}{a^2}\right)\frac{\partial C}{\partial x} \qquad (8\text{-}8)$$

In most cases radial diffusion predominates over axial diffusion. Thus, in most cases the term $\partial^2 C/\partial x^2$ may be neglected.

Now defining the dimensionless variables as

$$\bar{x} = \frac{x}{a}, \quad \bar{r} = \frac{r}{a}, \quad \tau = \frac{Dt}{\rho a^2} \quad \text{and} \quad \bar{C} = \frac{C}{C_0} \qquad (8\text{-}9)$$

where C_0 is a reference concentration, yields

$$\frac{\partial^2 \bar{C}}{\partial \bar{r}^2} + \frac{1}{\bar{r}}\frac{\partial \bar{C}}{\partial \bar{r}} \approx \frac{\partial \bar{C}}{\partial \tau} + 2\frac{a\rho\bar{v}}{D}(1 - \bar{r}^2)\frac{\partial \bar{C}}{\partial \bar{x}} \qquad (8\text{-}10)$$

where axial diffusion is neglected. The boundary condition at the wall of the tube

$$\frac{\partial \bar{C}}{\partial \bar{r}} = 0, \quad \text{at} \quad \bar{r} = 1 \qquad (8\text{-}11)$$

assures no diffusion through the wall.

An exact analytical solution of equation (8-8) would be most difficult to obtain. Taylor[13] has obtained approximate solutions which apply in certain cases. Following Taylor, we deduce the condition for which radial diffusion is more significant than axial convection.

A solution of equation (8-8) for which $\partial C/\partial x$ is zero is

$$\bar{C} = e^{-\alpha \tau} J_0(\sqrt{\alpha}\bar{r}) \qquad (8\text{-}12)$$

where J_0 is the Bessel function of the first kind of order zero. The boundary conditions (8-11) give

$$J_1(\sqrt{\alpha}) = 0 \qquad (8\text{-}13)$$

The root of this equation corresponding to the lowest value of α is $\sqrt{\alpha} = 3.8$. Thus the time required for the radial variation of $\bar{C}$ to decay to $1/e$ of its initial value is

$$\tau_1 = (3.8)^{-2} \quad \text{or} \quad t_1 = \frac{a^2 \rho}{D} (3.8)^{-2} \qquad (8\text{-}14)$$

On the other hand if the injected fluid is dispersed over a length l of the tube the time required for convection to make an appreciable change in $\bar{C}$ is of the order of

$$t_2 \approx \frac{2l}{\bar{v}} \qquad (8\text{-}15)$$

Thus, for convection to dominate over radial diffusion

$$t_2 \ll t_1 \qquad (8\text{-}16)$$

or

$$\frac{2l}{\bar{v}} \ll \frac{a^2 \rho}{D} (3.8)^{-2} \qquad (8\text{-}17)$$

Conversely, when

$$\frac{2l}{\bar{v}} \gg \frac{a^2 \rho}{D} (3.8)^{-2} \qquad (8\text{-}18)$$

radial diffusion will dominate over axial convection.

Taylor has obtained approximate solutions to equation (8-8) corresponding to these two extreme cases: Case 1 in which axial convection dominates the process, and case 2 in which radial diffusion dominates over axial convection. The solution of case 1 is as given by equation (8-7).

Case 2, in which radial diffusion dominates is of more interest in connection with displacement in porous materials. For displacing fluid injected uniformly at $x = 0$ starting at $t = 0$, Taylor's solution to equation (8-10) is

$$\tilde{C} = \begin{cases} \tilde{C}_0 \left[\dfrac{1}{2} - \dfrac{1}{2} \operatorname{erf} \left(\dfrac{\bar{x} - \beta\tau}{2\sqrt{\dfrac{\beta^2\tau}{48}}} \right) \right], & \bar{x} - \beta\tau > 0 \\[6mm] \tilde{C}_0 \left[\dfrac{1}{2} + \dfrac{1}{2} \operatorname{erf} \left(\dfrac{\beta\tau - \bar{x}}{2\sqrt{\dfrac{\beta^2\tau}{48}}} \right) \right], & \bar{x} - \beta\tau < 0 \end{cases} \qquad (8\text{-}19)$$

Here $\beta = a\rho\bar{v}/D$ is a dimensionless velocity factor and $\tilde{C}$ is the average value of $\tilde{C}$ over the cross section at x. The error function, erf z, is defined by

$$\operatorname{erf} z = \frac{2}{\sqrt{\pi}} \int_0^z e^{-\xi^2} \, d\xi \qquad (8\text{-}20)$$

and can be found in tables. In equations (8-19) $\tilde{C}_0$ is the concentration of the material being considered in the injected fluid at $x = 0$.

This solution shows that the displacement gives rise to a concentration distribution such as is shown in Figure 8-1. This result is confirmed by experiment for a rather wide range of values of $\beta = a\rho\bar{v}/D$. Aris[1] has been able to show that Taylor's solution for negligible axial diffusion also applies

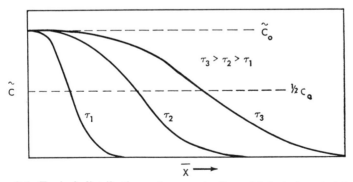

Figure 8–1. Typical distributions of concentration of injected material during linear miscible displacement as predicted by Taylor's solution.

when axial and radial diffusion are comparable in effect if our β in the radical above is replaced by another factor. Thus

$$\sqrt{\frac{\beta^2 \tau}{48}} \rightarrow \sqrt{(1 + \delta\beta^2)\tau} \qquad (8\text{-}21)$$

where $\delta = \frac{1}{48}$ for a circular tube and δ may have other values for a different geometry of cross section.

In terms of the original dimensional variables the argument of the error function is then

$$\frac{x - \bar{v}t}{2\sqrt{K_D t}}$$

where

$$K_D = D' + \delta \frac{a^2 \bar{v}^2}{D'} \qquad (8\text{-}22)$$

is the effective diffusion coefficient expressed in units of length squared per unit time (cm^2/sec). This factor is usually referred to as the dispersion coefficient. Note that for zero fluid velocity ($\bar{v} = 0$) K_D is just the molecular diffusion coefficient.

Aris was also able to show that δ is restricted to a rather narrow range of values. For elliptical cross sections, $\frac{5}{12} \leq 48\delta \leq 1$.

The mathematical solution given can be used to construct an operational definition of the effective dispersion coefficient. Such a definition allows direct measurement of K_D even when D', a and $\bar{v}$ are unknown.

From the above solution the distance, L, between the points at which $\tilde{C} = 0.10 \tilde{C}_0$ and $\tilde{C} = 0.90 \tilde{C}_0$ (the 10% and 90% concentration planes) is

$$L = 3.62 \sqrt{K_D t} \qquad (8\text{-}23)$$

where t is the time elapsed from the start of injection. L is called the length of the transition zone. Thus

$$K_D = \frac{L^2}{(3.62)^2 t} \qquad (8\text{-}24)$$

Consequently, if in a displacement experiment the distribution of $\tilde{C}$ is measured at some time t the value of K_D can be computed.

This discussion applies only for fluids of equal densities and viscosities. However, Blackwell[3] has shown from experiments that for fluids of unequal densities and viscosities the concentration distribution can be well approximated by the same error function solution. In particular, equation (8-24) was used to compute values of K_D for such systems. These values of K_D

indicated apparent values of D' different from the actual values for simple diffusion. These apparent or effective values of D' could be represented approximately as arithmetic averages of the two diffusion constants of the fluids.

8.30: Miscible Displacement in Porous Media

Miscible displacement of one fluid by another within a porous material is a potentially important process in the recovery of petroleum. Secondary recovery operations by waterflooding result in residual oil saturations within the reservoir rock which are relatively high. This is a result of the interfacial tension between water and oil. Flooding with a solvent bank should be much more efficient because the displacing fluid is completely miscible with the oil. Here some of the microscopic features of such displacement processes are considered.

The discussion of miscible displacement within a straight capillary tube given in the previous section is a guide to understanding certain features of miscible displacement in porous medial. Two factors are seen to be important on the microscopic scale. First, diffusion tends to create a dispersion of the front in the direction of flow. Second, convection creates dispersion in the direction of flow but this is accompanied by lateral diffusion within the pores.

Though on the macroscopic scale a linear displacement process is treated in terms of one-dimensional flow, this does not hold on the microscopic scale. In Chapter 7 the linear displacement problem was treated as one-dimensional flow. The macroscopic streamlines were then all parallel straight lines. These streamlines can only represent the average paths of fluid particles. On the microscopic scale a fluid element moves along a randomly tortuous path. The average displacement from the straight path is, of course, zero.

These tortuous paths of flow also are not uniform in cross section, and though two paths cannot cross, if the flow is laminar, they may pass through a common pore opening. Thus, mixing by diffusion between adjacent tubes of flow will occur.

Darcy's law and the continuity equation are combined to give a description of the macroscopic flow geometry. In order to construct a theory describing the dispersion associated with miscible displacement in porous media the problem must be studied on a microscopic scale as well. The purpose of such study is, of course, to construct a macroscopic description of the dispersion.

Experimental studies of the dispersion of an injected fluid when displacing another, miscible, fluid from a porous medium have been carried out by several investigators.[2, 4, 8, 16] Under conditions of linear laminar flow one would expect the concentration distribution (the average concentration in a plane perpendicular to the direction of flow as a function of distance and time) to be very similar to that observed in capillary tubes and this is actually observed.

Linear miscible displacement within a homogeneous porous medium gives rise to a concentration distribution which can be approximated by

$$\tilde{C} = \begin{cases} \tilde{C}_0[\tfrac{1}{2} - \tfrac{1}{2} \operatorname{erf} \gamma], & x - ut > 0 \\ \tilde{C}_0[\tfrac{1}{2} + \tfrac{1}{2} \operatorname{erf} \gamma], & x - ut < 0 \end{cases} \qquad (8\text{-}25)$$

for fluids of equal density and viscosity. Here $\tilde{C}$ is the average concentration in the plane a distance x from the injection face at time t and $\tilde{C}_0$ is the concentration in the injected fluid. u is the "mean pore velocity" given by

$$u = \frac{q}{\phi A} \qquad (8\text{-}26)$$

with q being the volumetric injection rate, ϕ the porosity and A the cross-sectional area of the sample. The argument, γ, of the error function is

$$\gamma = \frac{x - ut}{2\sqrt{K_D t}} \qquad (8\text{-}27)$$

Data from such displacement experiments can be used to compute effective dispersion coefficients, K_D, by using equation (8-24) of the previous section. Since for straight capillary tubes K_L/D' is a linear function of $(a\bar{v}/D')^2$ one would expect a similar relation to hold for porous media. However, it would be necessary to replace $\bar{v}$ by u, as defined in equation (8-26) and some estimate of a mean pore radius would have to replace the capillary radius, a. For unconsolidated sands a is usually replaced by the average particle radius, a_p, for correlation purposes.

Figure 8-2 illustrates the observed dependence of K_D/D' on $a_p u/D$ in unconsolidated sands. Here both scales are logarithmic scales. The data of Rafai,[8] which do not fall on the same curve with the other data, are for a porous material of a different texture than the other data, indicating an effect of texture.

Two important features of these data are to be noted. First of all the slope of the straight-line portion differs from the theoretical value of two. The slope here is 1.17. This indicates a type of convective dispersion in porous materials which differs considerably from that in a straight capillary tube. The second important feature is that in porous materials the limiting value of K_L/D' for small values of $a_p u/D'$ is not unity as in a straight capil-

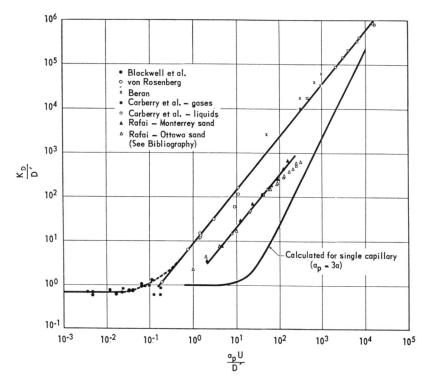

Figure 8–2. Dispersion coefficients for porous media showing the effect of displacement velocity. (*After Blackwell et al., 1958.*)

lary tube. This is readily accounted for as a result of the increased length of path for diffusion in porous materials. For pure diffusion ($u = 0$) in porous materials $K_D = D'/\tau$, where τ is the tortuosity of the medium. For unconsolidated sands $\tau \approx 1.5$ and this agrees with the limiting value, $K_D/D' \to 0.66$ for $u \to 0$ indicated in Figure 8-2.

8.31 Theory of Dispersion in Porous Media

The experimental results cited in the last section indicate the necessity of a theory of dispersion in porous media which takes account of the random microscopic nature of the porous structure. Thus the motion of a particular fluid element is to be viewed as a random process with little or no correlation between elements or space-time points. In the case of two miscible fluids having equal density and viscosity the flow problem is just that of a single homogeneous fluid. The dispersion problem is then viewed as that of marked (colored) fluid displacing unmarked (clear) fluid.

Several investigators[9, 10, 11] have undertaken to construct a mathematical theory describing this random process. Scheidegger[11] and also Saffman[9] have treated the problem in terms of a random walk formulation in analogy to Einstein's theory of Brownian motion. Saffman[10] has also treated the problem by the method of Lagrangian correlation functions.

Scheidegger's early treatment by the random walk formulation can be described as follows.

The porous medium is considered isotropic and macroscopically homogeneous. The external forces on the fluid are homogeneous and time independent, that is, a constant macroscopic pressure gradient in a unique direction. It is assumed that the average displacement of any fluid element corresponds to the macroscopic description afforded by Darcy's law.

The time interval 0 to t is divided into equal intervals T such that

$$NT = t \tag{8-28}$$

In every interval T the fluid element under consideration will undergo a displacement. The displacement from the mean (macroscopic) path during any T is viewed as a random process. Thus, the fluid element executes a random walk in the space $(x - \bar{x})$, $(y - \bar{y})$, $(z - \bar{z})$, where $\bar{x}, \bar{y}, \bar{z}$ are the coordinates of the mean or macroscopic displacement. Since the medium is homogeneous in the large, and isotropic, the probabilities for steps in all directions are equal and constant for every time interval. Scheidegger then shows that after many steps, large N, the probability for the fluid element to be at x, y, z after N steps will be

$$dP = (2\pi N\sigma^2)^{-3/2} \exp \left\{ -\frac{[(x - \bar{x})^2 + (y - \bar{y})^2 + (y - z)^2]}{2N\sigma^2} \right\} dxdydz \tag{8-29}$$

no matter what the initial distribution of probability may have been.

The quantities σ and T are constants during the motion and one may put

$$\sigma^2 = 2K_D T \tag{8-30}$$

where K_D is a dispersion factor to be determined in terms of other parameters. Both K_D and σ are independent of x, y, z and t but may depend on $\bar{x}, \bar{y}$ and $\bar{z}$.

Since $t = NT$, the expression for dP can be written as

$$dP = (4\pi K_D t)^{-3/2} \exp \left\{ -\frac{[(x - \bar{x})^2 + (y - \bar{y})^2 + (z - \bar{z})^2]}{4K_L t} \right\} dxdydz \tag{8-31}$$

Deviating now somewhat from Scheidegger's treatment it can be argued that the mean displacement $\bar{x}, \bar{y}, \bar{z}$ must be given by Darcy's law. Thus

$$\bar{x} = -\frac{tK}{\phi\mu}\frac{\partial p}{\partial x} = u_x t \qquad (8\text{-}32)$$

for the mean displacement in the x-direction, and $\bar{z}$ and $\bar{y}$ are zero.

Also since dP is the probability for a particle of fluid to be in the volume element $dxdydz$ at x, y, z at time t, it follows from the law of large numbers[15] that the actual number of marked particles in this volume element will be numerically equal to dP if a great number of particles is considered. (Note: for marked particles introduced continuously equation (8-31) must be integrated with respect to t.)

The probability dP is, due to the isotropic nature of the medium, the product of three independent probability distributions. For example

$$dp_x = (4\pi K_D t)^{-1/2} \exp\left\{-\frac{(x-\bar{x})^2}{4K_L t}\right\} dx \qquad (8\text{-}33)$$

is the probability for a marked particle to be between x and $x + dx$ at time t.

Since pure laminar flow is assumed there can be no interchange of particles between adjacent streamlines. Thus, for example, dP_x is a function of x and $\bar{x}$ only; it cannot depend on the velocity components, u_x for example. Using equation (8-32) to eliminate t from equation (8-33) shows that dP_x is a function of x and $\bar{x}$ only if

$$K_D = \alpha u_x \qquad (8\text{-}34)$$

where α is a constant characteristic of the porous medium.

This shows that dispersion arising solely from the tortuous pathways available to a fluid particle in a random porous structure corresponds to a dispersion coefficient proportional to the first power of the mean pore velocity u_x. This agrees rather well with the experimental results presented in the last section. The data presented there indicated the exponent to be 1.17 which is very close to unity.

Scheidegger's theory does not include molecular diffusion. In essence this corresponds to a uniform concentration over the cross section of each tortuous capillary pathway within the porous medium.

Scheidegger gives another evaluation of K_D based on the mechanics of flow within a capillary. However, his argument for this case is not so straightforward as that given above. This alternate treatment leads to a dependence of K_D on the square of the velocity.

Saffman[9, 10] has published two treatments of the dispersion problem. His first treatment, also based on the random walk analysis, leads to a dependence of K_D on the first power of the velocity. However, his second treatment,[10] based on Lagrangian correlation functions, yields a relation between K_D, D' and u which except for the factor δ is the same as that of straight capillary tubes. (Equation 8-22.)

Saffman also introduces the concept of a lateral dispersion coefficient. Such dispersion would arise between two fluids flowing parallel to each other in a porous medium both due to diffusion and the tortuous pathways of fluid elements.

In summary, a completely successful theory of dispersion in porous media has not been formulated, although the work of both Scheidegger and Saffman appears promising.

8.40: Macroscopic Features of Miscible Displacement

When miscible displacement is viewed from the macroscopic point of view the dispersion phenomena discussed in the two previous sections are not of such great importance. This is particularly true of miscible displacement on the scale of petroleum reservoirs.

The mean pore velocity in secondary recovery operations in petroleum reservoirs rarely exceeds a value of about one foot per day, or about 3.5×10^{-4} cm/sec. Then, for a typical diffusion constant of 2×10^{-5} cm^2/sec and a value of a_p of the order of 0.02 cm the group $a_p u / D'$ has a value of the order of 0.35. Hence from the results of Blackwell *et al* presented in Figure 8-2 the dispersion factor, K_D, is just D'/τ. As a result the dispersion factor can be treated as independent of flow rate, that is, only ordinary diffusion, modified by the path length factor, τ, need be considered. This conclusion must be modified in the neighborhood of wells. Here the fluid velocity is quite high and consequently the dispersion factor is greater than D'/τ.

These conclusions regarding K_D in reservoir problems also should hold for fluids which differ in viscosity and density. This follows from results obtained by Blackwell[3] for displacements in capillary tubes with such fluids.

Because K_D can be represented by D'/τ at reservoir flow rates mass transfer in the reservoir can be approximated by two independent processes, ordinary diffusion as governed by Fick's law with D' replaced by D'/τ and mass transport due to flow. Thus the rate of transport per unit area, expressed as the mass flux density in molecules per unit time per unit area, is

$$\hat{N} = -\frac{D'}{\tau} \nabla C + C\hat{v} \qquad (8\text{-}35)$$

Here $\hat{v}$ is the volume flux density as given by Darcy's law (see section 3.30)

$$\hat{v} = - \frac{K}{\mu} (\nabla p + \rho g \hat{\imath}_3) \tag{8-36}$$

The unit vector is directed upward along the positive x_3 axis. In this equation μ and ρ are local values which may depend on the composition of the fluid. In equation (8-35) the concentration is expressed in molecules per unit volume and is an average value over several pores.

Requiring conservation of molecules of the substance in question, we apply the continuity equation. Thus we obtain

$$\frac{D'}{\tau} \nabla^2 C - \nabla \cdot (\hat{v} C) = \phi \frac{\partial C}{\partial t} \tag{8-37}$$

Also, if the fluids are incompressible and the fluid density is a function of composition, $\rho = \rho(C)$, the continuity equation for conservation of fluid mass combined with Darcy's law yields

$$\nabla \cdot \left[\frac{k\rho}{\mu} (\nabla p + \rho g \hat{\imath}_3) \right] = \phi \frac{d\rho}{dC} \frac{\partial C}{\partial t} \tag{8-38}$$

These two equations, coupled with the equation

$$\mu = \mu(C) \tag{8-39}$$

which relates the viscosity as a function of composition, describe the miscible displacement process. Of course, some simplifications have been introduced. Recently numerical solutions of these equations have been obtained by Douglas, Peaceman and Rachford.[5] However, they included a lateral dispersion factor also. Their results for a two-dimensional system with a random spatial distribution of permeability show exactly the same kind of fingering behavior observed in laboratory models (see Figure 7-13).

Now the relative importance of diffusion and transport on the macroscopic scale can be estimated. Mass transport by diffusion is of the order

$$\dot{N}_D \approx \frac{D'}{\tau} \frac{C}{\Delta l} \tag{8-40}$$

while convective transport is of the order of

$$\dot{N}_T \approx Cv \tag{8-41}$$

The ratio of these two rates is

$$\frac{\dot{N}_T}{\dot{N}_D} \approx \frac{v \tau \Delta l}{D'} \tag{8-42}$$

Here Δl is the order of magnitude of the length of the transition zone across which appreciable concentration gradients exist. Thus Δl can be estimated by equation (8-23). In this case

$$\Delta l \approx 3.62 \sqrt{\frac{D'}{\tau} t} \tag{8-43}$$

and therefore

$$\frac{\dot{N}_T}{\dot{N}_D} \approx 3.62 \, v \sqrt{\frac{\tau}{D'} t} \tag{8-44}$$

For flooding times of the order of years this yields, for typical values of v and D', the magnitude 10^5. Hence transport by flow dominates over diffusion, and when compared to reservoir dimensions the transition zone will be quite small, i.e. Δl is given by equation (8-43) and the distance, l, traveled by the front is of the order of ut. For $t \approx 1$ year this gives $\Delta l/l \approx 10^{-2}$.

As a consequence of the above result the displacement can be treated approximately as a frontal process as discussed in Chapter 7. Behind the front c, μ and ρ have one set of uniform values and ahead of the front a different set of uniform values.

From this point of view it is evident that the phenomenon of viscous fingering (section 7-7) will be of importance if the viscosity of the invading fluid is less than that of the displaced fluid. Also due to differences in fluid density gravity segregation will be important. The invading fluid will tend either to under-run the resident fluid or to over-ride the resident fluid. These macroscopic phenomena are the principal factors to be considered in miscible displacement processes in petroleum reservoirs.

Under certain conditions, extreme fingering or channeling through high permeability streaks and severe gravity segregation, diffusion may become of importance in another way. The displacing fluid is flowing past essentially stationary resident fluid. The only exchange between the fluids is by diffusion.

The extreme complexity of the interplay of dispersion, diffusion and flow in miscible displacement makes the mathematical treatment of such problems all but impossible except in the simplest of cases. Consequently, models are widely used for the study of such problems.

8.50: Turbulent Flow of Miscible Fluids

Only laminar flow of miscible fluids has been considered in this chapter. Very little work has been done on the turbulent flow of miscible fluids through porous media. Taylor[14] has developed a rather successful theory of turbulent flow of miscible fluids in pipes. Very likely this theory can be extended by analogy to porous media in much the same way as has been done for the laminar case.

An adequate theory of turbulent miscible displacement in porous media would have important applications in the treatment of gas-liquid chromatography.

EXERCISES

1. Considering no diffusion show that displacement of one fluid by another in a circular capillary tube yields the average concentration of displacing fluid in a cross section as:

$$
C\ (x,\ t)\ =\ \begin{cases} C_0 \dfrac{2\bar{v}t\ -\ x}{2\bar{v}t}, & x\ <\ 2\bar{v}t \\[3mm] 0 & x\ >\ 2\bar{v}t \end{cases}
$$

Here C_0 is concentration of substance in injected fluid at $x = 0$.

2. Consider a porous medium as a bundle of parallel circular capillary tubes having some distribution of radius. For each tube $\tilde{C}(x,\ t)$ as given in Problem 1 describes the average concentration in the cross section with $\bar{v} = r^2 \Delta p / 8 \mu L$. Here $\Delta p =$ pressure drop across tube of length, L. Write the equation for the average concentration in a cross section of the porous medium if

$$
\frac{dN}{N}\ =\ \frac{1}{2\alpha^2}\ re^{-\alpha r^2}\ dr
$$

is the fraction of tubes with radius between r and $r + dr$.

References

1. Aris, R., *Proc. Roy. Soc. (London)*, A235, 67 (1956).
2. Beran, M. J., "Dispersion of Soluable Matter in Slowly Moving Fluid," Dissertation, Harvard Univ. (1955).
3. Blackwell, R. J., "An Investigation of Miscible Displacement Processes in Capillaries," presented at local sections, *Am. Inst. Chem. Engrs.*, Galveston, Texas, Oct. (1957).
4. Blackwell, R. J., Rayne, J. R., and Terry, W. M., *Trans. Am. Inst. Mining Met. Engrs.*, **217**, 1 (1958).
5. Carberry, J. J., and Bretton, R. H., "Axial Dispersion in Flow Through Fixed Beds," presented at Am. Inst. Chem. Engrs. meeting, Chicago (1957).
6. Douglas, J., Jr., Peaceman, D. W., and Rachford, H. H., Jr., Private communication (1960).
7. Glasstone, S., "Textbook Of Physical Chemistry," D. Van Nostrand, New York (1946).
8. Rafai, M. N. E., "An Investigation of Dispersion Phenomena in Laminar Flow through Porous Media," Dissertation, Univ. of Calif. (1956).
9. Saffman, P. G., *J. Fl. Mech.* **6**, 321, (1959).
10. Saffman, P. G., *J. Fl. Mech.*, **7**, 194 (1960).

11. Schiedegger, A. E., *J. Appl. Phy.* **25,** 994 (1954).
12. Sommerfeld, A., "Mechanics of Deformable Bodies," Academic Press, Inc., New York (1950).
13. Taylor, G., *Proc. Roy. Soc.* (*London*), **A219,** 186 (1953).
14. Taylor, G., *Proc. Roy. Soc.* (*London*), **A223,** 446 (1954).
15. Uspensky, J. V., "Introduction to Mathematical Probability," McGraw-Hill Book Co., New York (1937).
16. Von Rosenberg, D. U., *Am. Inst. Chem. Engrs. J.*, **2,** 55 (1956).

9. THEORY OF MODELS

9.10: The Concept of Similarity

The theory of models is based on the concept of similarity. In plane geometry the concept of similarity is employed in discussing such things as "similar triangles." It is usually stated that for similar triangles the ratio of two sides in one triangle is equal to the corresponding ratio in another triangle. This property of similar triangles is independent of the size of the triangles. In order to give a clear definition of this concept of similarity, some basic concepts of projective geometry must be employed. Only the projection of a plane needs to be considered.

Suppose, as is depicted in Figure 9-1, a geometrical figure is given on a plane, S, and a point P, not in the plane S, is arbitrarily selected. Then, a straight line can be constructed from each point of the given figure to the point, P. If, then, a plane, S', parallel to S, is placed between S and P, each of these straight lines will intersect S' at only one point. The locus of all such points defines a geometrical figure on S'. The figure so constructed is defined as being similar to the given figure.

This concept of geometrical similarity can be stated in the following way. Two plane figures are defined as being geometrically similar if one is the point projection of the other on a parallel plane. The most important aspect of this definition is that it implies equality of all properties of the figures which are independent of absolute size.

The generalization of this definition of similarity to more than two dimensions can be accomplished but need not be undertaken here.

Dynamic and kinematic similarity are concepts which imply equality between certain variables in "similar" physical systems. Such similarity can be put on the same basis as geometrical similarity. Suppose the dynamic and/or kinematic behavior of a physical system can be described by plotting a variable y versus a variable x. Since y and x represent measurable properties of the system, some unit of measurement must be used in each case. Thus, when plotted on the same scale, it may be that the shape of the curve would be altered by using a different unit of measurement for, say, the x variable. However, if a different unit for the y variable were also employed, the shape of the curve might not be altered. It is obvious that, for linear scales, if the y unit and the x unit are both doubled the shape of the curve would not be altered. The curve with doubled units would be simply

217

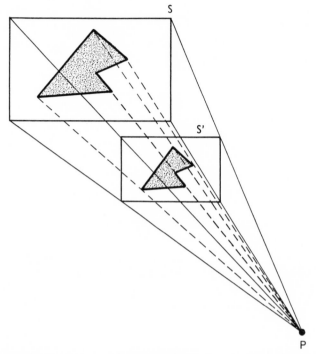

Figure 9–1. The concept of geometrical similarity; the planes are parallel.

a plane projection of the curve with single units. Consequently, the two curves would be similar in the same sense that two geometrical figures may be similar.

While it is possible to proceed along this line of reasoning to construct a logical basis for models in a geometrical form, the analytical approach is much more efficient. This is undertaken in the following section.

9.20: Basic Theory of Models

A general theory of models can be constructed without any reference to a particular physical system. To accomplish this, let it be supposed that all aspects of a physical system can be described completely by a system of mathematical equations, E. This means that the physical system is assumed to be a causal system and completely determinant. This requires that a given initial state of the system leads to a unique sequence of subsequent states. The requirement that the system of equations E be complete simply means that all variables and parameters must be included to ensure the correct predication of all states of the physical system. This can be stated symbolically as follows.

Let the dependent variables of the system be denoted by y_1, $y_2 \cdots y_m$; the independent variables by x_1, $x_2 \cdots x_n$; and the parameters by a_1, a_2, $\cdots a_k$. Then the system of equations E can be written symbolically as

$$E\{y_1, y_2, \cdots y_m ; x_1, x_2, \cdots x_n ; a_1, a_2, \cdots a_k\} \qquad (9\text{-}1)$$

This system may be composed of several types of equations, integral, differential, or some functional type, plus boundary conditions or other equations of constraint. The requirement of completeness simply means that the system E must possess unique solutions giving the y_i as functions of the x_j and the parameters a_s. Furthermore, the requirement that the system of equations describes the physicial system means that these solutions must correspond to actual states of the physical system. These solutions can be denoted by

$$y_i = y_i(x_1, x_2, \cdots x_n ; a_1, a_2, \cdots a_k), \qquad i = 1, 2, \cdots m \quad (9\text{-}2)$$

The number of variables and parameters included in these solutions may, in general, be less than the number included in the system E. This means that, in some cases, the y_i may be independent of some of the variables, x_j, or parameters, a_s. If the number of variables and parameters included in the solutions is denoted by N, then

$$N \leq m + n + k \qquad (9\text{-}3)$$

where m is the number of y_i, n is the number of x_j, and k is the number of a_s included in the system of equations, E.

Since the system E and its solutions are to describe measurable properties of a physical system, all the equations involved must be dimensionally homogeneous. If there are D_s independent dimensions[1] included in equation (9-2), then by the Buckingham π-Theorem* the N dimensional quantities included in the solutions can be combined into $\bar{N}_s$ dimensionless groups, where

$$\bar{N}_s = N - D_s \qquad (9\text{-}4)$$

such that the solutions are expressed in dimensionless form in terms of only these $\bar{N}_s$ dimensionless quantities.

Similarly, the system E can be transformed to a dimensionless system, E', expressed in terms of $\bar{N}_E$ dimensionless quantities where

$$\bar{N}_E = (n + m + k) - D_E \qquad (9\text{-}5)$$

and D_E is the number of independent dimensions included in the system E. From equations (9-3) and (9-4) and (9-5) it follows that

$$\bar{N}_s + D_s \leq \bar{N}_E + D_E \qquad (9\text{-}6)$$

---

* See appendix at end of chapter.

Hence, if the solutions of the system E contain the same number of independent dimensions as E itself, then $\bar{N}_s$ is either less than or equal to $\bar{N}_E$.

Usually a more restricted procedure is employed for constructing a dimensionless system, E', from the system, E, than indicated above. The method employed most frequently in the design of scaled models is to use a simple change of variables, or set of transformations, which transforms the dimensional variables in E to dimensionless variables. In this manner a completely dimensionless system, E', is obtained, since for dimensional homogeneity of the equations the parameters occurring in E will automatically be transformed to the dimensionless parameters of E' by these transformations. Thus, let

$$\left.\begin{array}{l} Y_i = Y_i(y_i, a_1, a_2, \cdots a_k) \quad i = 1, 2, \cdots m \\ X_j = X_j(x_j, a_1, a_2, \cdots a_k) \quad j = 1, 2, \cdots n \end{array}\right\} \tag{9-7}$$

be the set of transformations from the x_j, y_i to the dimensionless X_j, Y_i. These yield, when applied to E, the dimensionless system

$$E'\{Y_1, Y_2, \cdots Y_m; X_1, X_2, \cdots X_n; A_1, A_2 \cdots A_{k'}\} \tag{9-8}$$

In this case the number of Y_i is the same as the number y_i and similarly for the x_j and X_j, but the number k', of dimensionless parameters A_s, is not the same as the number, k, of the a_s'. In fact, it follows from the π-Theorem that k and k' are related by

$$k' = k - D_E \tag{9-9}$$

That is, the number of dimensionless parameters in the dimensionless system, E', is less than the number of parameters in the dimensional system, E, by D_E, where D_E is the number of independent dimensions in E.

The important distinction between transforming the solutions (equation 9-2) to dimensionless form, and transforming the system E itself to dimensionless form lies in the fact that the solutions may contain fewer variables and parameters than the system E. The significance of this will be made more evident in the following paragraphs.

Considered now are the solutions of the dimensionless system, E', which are written

$$Y_i = Y_i(X_1, X_2, \cdots X_n; A_1, A_2, \cdots A_{k'}) \quad i = 1, 2, \cdots m \tag{9-10}$$

It is noted that, in general, a family of solutions exists, each of which is characterized by a particular set of values of $A_1, A_2, \cdots, A_{k'}$. This means that two members of this family of solutions, the Y_i as functions of the X_j, are identical if, and only if, each A_s has the same numerical value in both solutions.

9.30: Scaled Models and Scaling Laws

Let two physical systems be considered, one with a system of defining equations, E, and another with defining equations, G. Furthermore, it is assumed that for each equation in E there is a correpsonding equation of exactly the same form in G (i.e., for each symbol and operation in E there is a corresponding symbol and the same operation in G). When these conditions are satisfied, one system is said to be the analog of the other. It is not necessary that the physical dimensions in the two systems be the same. When the physical dimensions of the two systems are the same, then one is called simply a model of the other instead of an analog model.

Now, let it be supposed that by a set of transformations of the form of equations (9-7), the E system is transformed to a dimensionless system E', and by a similar set of transformations, G is transformed to a dimensionless system $G.'$ The two systems, E' and G', are identical in form, which means that the family of solutions of E' corresponding to different sets of values of its dimensionless parameters is identical to the family of solutions of G' corresponding to different sets of values of its dimensionless parameters. Thus, if the parameters of E' are denoted by A_{Es}, $s = 1, 2, \cdots, k'$, and the parameters of G' by A_{Gs}, $s = 1, 2, \cdots k'$, a solution of E' will be numerically identical to a solution of G' if

$$A_{Es} = A_{Gs}, \qquad s = 1, 2, \cdots k' \quad (9\text{-}11)$$

These are the scaling requirements for the model.

As mentioned previously, the solution of D' or G' may contain fewer parameters than the system itself. Thus, it can happen that among the k' dimensionless parameters one or more do not occur in the solutions, in which case these can be deleted in the scaling requirements. In practice, however, the system of equations E' or G' can rarely be solved analytically and, hence, knowledge of which parameters are important for scaling purposes can only be established by experiment. Thus, without any knowledge of the mathematical form of the solutions, all the dimensionless parameters of E' and G' must be included in the scaling requirements.

It cannot be emphasized too strongly that only those features of the physical system correctly described by the system of equations, E, are correctly represented by a model constructed on the basis outlined. Thus, this model will yield numerical values of the Y_i as functions of the X_j which correspond to those of the prototype system. Also, any quantities derivable from the Y_i and X_j by ordinary mathematical operations that are consistent with the scaling requirements of equation (9-11) will be correctly represented. For example, quantities such as

$$\frac{\partial Y_i}{\partial X_j}, \quad \int Y_i \, dX_j, \quad \text{etc.} \tag{9-12}$$

will be correctly represented by the model.

Scaled models based on an analysis such as outlined above always contain implicit assumptions as to the nature of the physical system under consideration. For example, in the flow of fluids through porous media Darcy's law is usually included in the system of equations, E, and also the system, G, if the model is not an analog. Hence, the implicit assumption is made that in both systems Darcy's law is obeyed. Such assumptions should be recognized.

Frequently, the nature of the physical system is not adequately known so that it is not possible to write out a complete system of equations, E, describing the physical system. In this case it is still possible to arrive at an adequate set of scaling requirements for a model (but not an analog model) if the nature of the system is understood to the extent that a complete list of all variables and parameters pertinent to the problem can be made. For example, let it be supposed that the list so constructed contains n variables and k parameters. From these, according to the π-Theorem, n' dimensionless variables and k' dimensionless parameters can be constructed, where $n' + k' = n + k - D$. Here, D is the number of independent physical dimensions occurring in the list. If a model system is constructed with similar geometry, etc., and the same numerical values of these k' parameters, then the n' dimensionless variables will assume the same numerical values in the model as in the prototype system.

It should be noted that an analog model can be designed only when the complete system of equations, E, is known. This is true because it is the form of the equations that is the basis of an analog.

With this discussion of the basic principles of models and scaling now essentially complete, it is in order to consider some particular examples.

9.40: Examples of Scaling Analysis When the Equations for the System are Known

As an example of scaling analysis, the displacement of oil by water in a homogeneous, isotropic, porous medium is now considered. The fluids will be assumed incompressible.

Let it be assumed that the system under consideration is a rectangular parallelopiped defined by

$$0 \leq x_j \leq L_j, \qquad j = 1, 2, 3 \tag{9-13}$$

Further, let it be assumed that at $x = 0$, water is injected at a constant rate per unit area, v, and along the lateral surface of the figure no flow crosses the boundaries. Let the initial water saturation be S_{wm}. The system of equations describing this flow regime is obtained from Darcy's law as

$$E \equiv \begin{cases} \sum_{i=1}^{3} \frac{1}{\phi\mu_w} \frac{\partial}{\partial x_i} \left[K_w \frac{\partial}{\partial x_i} (p_w + \rho_w g x_3) \right] = \frac{\partial S_w}{\partial t} \\[2mm] \sum_{i=1}^{3} \frac{1}{\phi\mu_o} \frac{\partial}{\partial x_i} \left[K_o \frac{\partial}{\partial x_i} (p_o + \rho_o g x_3) \right] = \frac{\partial S_o}{\partial t} \\[2mm] p_o = p_w + p_c \\[2mm] S_w + S_o = 1 \\[2mm] v = \text{constant at input face} \\[2mm] S_w = S_{wm} \text{, initially} \\[2mm] \text{no flow across lateral surfaces} \\[2mm] \text{plus boundary conditions on } S_w \\[2mm] 0 \leq x_i \leq L_i, \quad i = 1, 2, 3 \end{cases} \qquad (9\text{-}14)$$

Here the x_i, $i = 1, 2, 3$ are rectangular cartesian coordinates with x_3 being positive vertically upward.

The boundary conditions on p_w are adequately included here. Those for S_w can be expressed in terms of the relative permeability and capillary-pressure functions.

The system of equations given and implied in (9-14) constitutes the E system for this problem. In this system there are three independent physical dimensions. The dependent variables can be taken as p_w and S_w and the independent variables are x_1, x_2, x_3, and t. Parameters occurring here are v, $\phi\mu_w$, $\phi\mu_o$, $\rho_w g$, $\rho_o g$, L_1, L_2, L_3, S_{wm} and also an unknown number of parameters required to express K_o, K_w, and p_c as functions of S_w.

The functions K_o, K_w and p_c for different porous media exhibit great variations in form. This is shown in Figures 9-2 and 9-3. However, certain critical points are common to all such curves for every porous medium. These critical points are the connate water saturation, S_c, at which $K_w \to 0$ and $p_c \to \infty$, and the residual oil saturation, S_{ro}, at which $K_o \to 0$ and $p_c \to 0$ on the imbibition capillary-pressure curve. It is better to consider K_o, K_w and p_c as functions of a modified saturation function. Thus, instead of S_w we employ the saturation function

$$\bar{S}_w = \frac{S_w - S_c}{1 - S_c - S_{ro}} \qquad (9\text{-}15)$$

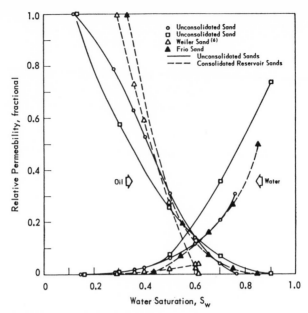

Figure 9–2. Samples of two-fluid permeability data. (*After Collins and Perkins, 1960.*)

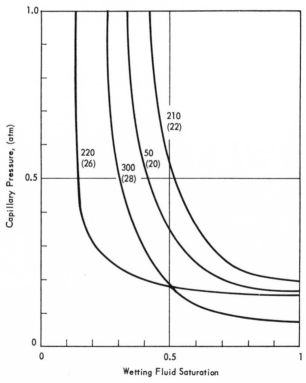

Figure 9–3. Samples of drainage capillary pressure data. Upper number on curves is permeability in millidarcies, lower number is porosity fraction.

and corresponding to S_o , we employ

$$\bar{S}_o = 1 - \bar{S}_w = \frac{S_o - S_{ro}}{1 - S_c - S_{ro}} \tag{9-16}$$

Also since at $S_w = S_c$ we have $K_o \rightarrow K_{oc}$ and at $S_o = S_{ro}$ we have $K_w \rightarrow$ K_{wr} , we define the relative permeabilities as

$$k_w = \frac{K_w}{K_{wr}} \tag{9-17}$$

and

$$k_o = \frac{K_o}{K_{ro}} \tag{9-18}$$

Note that these differ from the relative permeabilities defined elsewhere in this volume and generally employed in the petroleum industry. These are to be considered as functions of $\bar{S}_w$ as defined above.

That these definitions of relative permeabilities introduce a certain degree of uniformity for the variety of porous media encountered in application is shown in Figure 9-4. These are the same data shown in Figure 9-2.

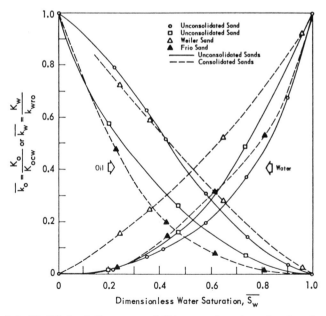

Figure 9-4. Modified relative permeability as a function of reduced saturation. (*After Collins and Perkins, 1960.*)

The capillary pressure curves are handled in a similar manner. We follow Leverett[6] by observing that the general level of p_c is determined by the interfacial tension, γ, a mean contact angle θ and a mean pore radius characterized by $(K/\phi)^{1/2}$. Thus define a dimensionless capillary pressure as

$$\bar{p}_c = \sqrt{\frac{K}{\phi}} \frac{p_c}{\gamma \cos \theta} \tag{9-19}$$

This function is also to be considered as a function of $\bar{S}_w$ during imbibition but should be a function of $\bar{S}_w' = (S_w - S_c)/(1 - S_c)$ for drainage.

The uniformity introduced to the capillary-pressure data of Figure 9-3 by this definition is shown in Figure 9-5. It should be noted that there

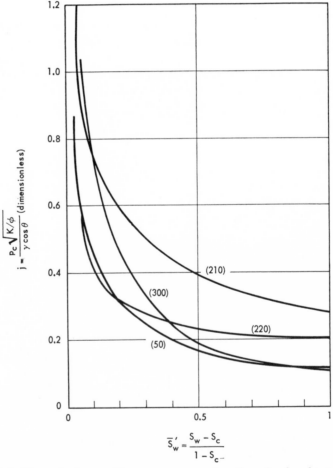

Figure 9–5. Dimensionless capillary pressure as a function of reduced saturation.

is always considerable uncertainty regarding the value of cos θ in this relationship.

The transformation of the system E above to a dimensionless E' system is accomplished by

$$\left.\begin{aligned} \bar{x}_i &= \frac{x_i}{L_i}, \qquad i = 1, 2, 3 \\[2mm] \tau &= \frac{vt}{\phi L_1(1 - S_{ro} - S_c)} \\[2mm] \bar{p}_w &= \frac{K_{wr}p_w}{v\mu_w L_1} \\[2mm] \bar{p}_o &= \frac{K_{wr}p_o}{v\mu_w L_1} \\[2mm] \bar{S}_w &= \frac{S_w - S_c}{1 - S_c - S_{ro}} \\[2mm] \bar{p}_c &= \sqrt{\frac{K}{\phi}} \frac{p_c}{\gamma \cos \theta} \end{aligned}\right\} \qquad (9\text{-}20)$$

The E' system thus obtained is

$$E' \equiv \begin{cases} \displaystyle\sum_{i=1}^{3} \left(\frac{L_1}{L_i}\right)^2 \frac{\partial}{\partial \bar{x}_i}\left[k_w \frac{\partial}{\partial \bar{x}_i}\left(\bar{p}_w + \frac{K_{wr}\rho_w g L_3}{v\mu_w L_1}\bar{x}_3\right)\right] = \dfrac{\partial \bar{S}_w}{\partial \tau} \\[4mm] \dfrac{\mu_w K_{oc}}{\mu_o K_{wr}} \displaystyle\sum_{i=1}^{3} \left(\frac{L_1}{L_i}\right)^2 \frac{\partial}{\partial \bar{x}_i}\left[k_o \frac{\partial}{\partial \bar{x}_i}\left(\bar{p}_o + \frac{K_{wr}\rho_o g L_3}{v\mu_w L_1}\bar{x}_3\right)\right] = \dfrac{\partial \bar{S}_o}{\partial \tau} \\[4mm] \bar{S}_w + \bar{S}_o = 1 \\[3mm] \bar{p}_o - \bar{p}_w = \dfrac{K_{wr}}{v\mu_w L_1}\sqrt{\dfrac{\phi}{K}} \, (\gamma \cos \theta) \, \bar{p}_c \\[3mm] \text{No flow across lateral boundaries} \\[2mm] \text{Unit dimensionless injection velocity} \\[2mm] \bar{S}_w = \dfrac{S_{wm} - S_c}{1 - S_c - S_{ro}} \quad \text{at} \quad \tau = 0 \\[3mm] \text{Boundary conditions of } \bar{S}_w \\[2mm] 0 \le \bar{x}_i \le 1, \quad i = 1, 2, 3 \end{cases} \qquad (9\text{-}21)$$

The dimensionless parameters included here are

$$\frac{L_1}{L_2}, \quad \frac{L_1}{L_3}, \quad \frac{K_{wr}\rho_w g L_3}{v\mu_w L_1}, \quad \frac{K_{wr}\rho_o g L_3}{v\mu_w L_1}, \quad \frac{\mu_w K_{oc}}{\mu_o K_{wr}}, \quad \frac{K_{wr}}{v\mu_w L_1}\sqrt{\frac{\phi}{K}}\,\gamma \cos \theta$$

and $(S_{wm} - S_c)/(1 - S_c - S_{ro})$. Thus according to the theory outlined above two flow systems of this type and of similar geometry will exhibit the same dependence of $\bar{p}_w$ and $\bar{S}_w$ on the $\bar{x}_i$ and τ, if these parameters have the same values for the two systems and k_w, k_o and $\bar{p}_c$ are the same functions of $\bar{S}_w$ in the two systems. Also the dimensionless fluid velocities will be the same functions of the $\bar{x}_i$ and τ in the two systems. It must be noted that the boundary conditions in the two systems must be the same (in dimensionless terms).

Since the entire analysis could have been carried out in terms of the flow potentials

$$\psi_w{}' = p_w + \rho_w g x_3 \quad \text{and} \quad \psi_o{}' = p_o + \rho_o g x_3 \qquad (9\text{-}22)$$

observe that the alternate E' system below can be obtained by defining

$$\bar{\psi}_w = \frac{K_{wr}\psi_w{}'}{v\mu_w L_1} \quad \text{and} \quad \bar{\psi}_o = \frac{K_{wr}\psi_o{}'}{v\mu_w L_1} \qquad (9\text{-}23)$$

Thus

$$E'' \equiv \begin{cases} \displaystyle\sum_{i=1}^{3} \left(\frac{L_1}{L_i}\right)^2 \frac{\partial}{\partial \bar{x}_i}\left(k_w \frac{\partial \bar{\psi}_w}{\partial \bar{x}_i}\right) = \frac{\partial \bar{S}_w}{\partial \tau} \\[2mm] \displaystyle\frac{\mu_w K_{oc}}{\mu_o K_{wr}} \sum_{i=1}^{3} \left(\frac{L_1}{L_i}\right)^2 \frac{\partial}{\partial \bar{x}_i}\left(k_o \frac{\partial \bar{\psi}_o}{\partial \bar{x}_i}\right) = \frac{\partial \bar{S}_o}{\partial \tau} \\[2mm] \bar{S}_o + \bar{S}_w = 1 \\[2mm] \bar{\psi}_w + \dfrac{K_{wr}}{v\mu_w L_1}\sqrt{\dfrac{\phi}{K}}\,\gamma\,\cos\,\phi\bar{p}_c - \dfrac{K_{wr}\Delta\rho g L_3}{v\mu_w L_1}\,\bar{x}_3 = \bar{\psi}_o \\[2mm] \text{No flow across lateral boundaries} \\[1mm] \text{Unit dimensionless injection velocity} \\[1mm] \bar{S}_w = \dfrac{S_{wm} - S_c}{1 - S_c - S_{ro}} \quad \text{at} \quad \tau = 0 \\[2mm] \text{boundary conditions on } \bar{S}_w \\[1mm] 0 \le \bar{x}_i \le 1 \quad i = 1, 2, 3 \end{cases} \qquad (9\text{-}24)$$

Here $\Delta\rho = \rho_w - \rho_o$. In this formulation $\bar{\psi}_w$ and $\bar{S}_w$ will be the same functions of the $\bar{x}_i$ and τ in two such systems having similar geometry and boundary conditions if the parameters

$$\frac{L_1}{L_2}, \quad \frac{L_1}{L_3}, \quad \frac{K_{wr}\Delta\rho g L_3}{v\mu_w L_1}, \quad \frac{\mu_w K_{oc}}{\mu_o K_{wr}}, \quad \frac{K_{wr}}{v\mu_w L_1}\sqrt{\frac{\phi}{K}}\,\gamma\,\cos\,\theta, \quad \frac{S_{wm} - S_c}{1 - S_{ro} - S_c}$$

have the same values in the two systems.

Here then we see that one less scaling group is required. Thus, the second formulation may be thought to be preferred because the scaling requirements are less stringent. However, in the first analysis both dimensionless flow rates and dimensionless pressures are scaled, while in the second analysis the dimensionless potentials and flow rates are scaled. Thus, when considering any set of scaling groups for a particular physical problem the question as to what variables of the system are comparable in model and prototype must be considered. For some studies the second analysis given would be satisfactory, but in other cases the first might be necessary.

The procedures described in this example can be applied to any physical problem for which the physical processes are well enough understood to permit complete mathematical formulation of the problem. Perhaps the greatest source of error in this method is an incorrect mathematical formulation of the physical problem.

9.50: Example of Analog Construction

As an example of analog construction, the theoretical basis for the potentiometric model of single-phase, steady-state flow in an isotropic porous medium will be considered. For simplicity, we treat the two-dimensional case.

Following the lines of development employed for steady plane flow employed in Chapter 4, we obtain

$$\frac{\partial}{\partial x}\left(\beta \frac{\partial U}{\partial x}\right) + \frac{\partial}{\partial y}\left(\beta \frac{\partial U}{\partial y}\right) = 0 \tag{9-25}$$

as the differential equation of flow. Here

$$U = \frac{MK_r}{2RT\mu} p(p + 2b) \tag{9-26}$$

if the fluid is a gas. The permeability of the medium, K, is variable from point to point so we define

$$\beta = \frac{K}{K_r} \tag{9-27}$$

where K_r is a reference value, the value at $x = y = 0$ for example. Here x and y are rectangular Cartesian coordinates in the plane.

If the fluid is an incompressible fluid, we let

$$U = \frac{K_r\rho}{\mu} (p + \rho g x_3) \tag{9-28}$$

again making reference to Chapter 4, section 4.31.

We note that

$$\dot{m} = -\beta\nabla U \tag{9-29}$$

is the mass flow rate per unit area in either case.

Now consider the flow of electric current through an isotropic conducting sheet of uniform thickness. From the conservation of charge and Ohm's law, we obtain

$$\frac{\partial}{\partial x}\left(\sigma\frac{\partial V}{\partial x}\right) + \frac{\partial}{\partial y}\left(\sigma\frac{\partial V}{\partial y}\right) = 0 \tag{9-30}$$

where V is the electric potential and σ is the conductivity of the medium which may be variable from point to point. Here

$$\hat{j} = -\sigma\nabla V \tag{9-31}$$

is the current density according to Ohm's law.

Now the equations apply only for uniform thickness of the region of the flow, both in the porous stratum and the conductor. However small variation in thickness in either case can be included in a rather simple way.

Variations in thickness introduce vertical components of flow. However, for small variations these will be small and the velocity vector will still lie approximately in a horizontal plane. In this case we have approximately

$$\frac{\partial}{\partial x}\left(\beta\frac{H}{H_r}\frac{\partial U}{\partial x}\right) + \frac{\partial}{\partial y}\left(\beta\frac{H}{H_r}\frac{\partial U}{\partial y}\right) = 0 \tag{9-32}$$

for the fluid-flow problem, and

$$\frac{\partial}{\partial x}\left(\sigma h\frac{\partial V}{\partial x}\right) + \frac{\partial}{\partial y}\left(\sigma h\frac{\partial V}{\partial y}\right) = 0 \tag{9-33}$$

for the electrical problem. Here $H(x, y)$ is the thickness of the porous stratum, H_r is a reference value, $H(0, 0)$ for example and $h(x, y)$ is the thickness of the conductor.

We can now write a dimensionless description of the fluid-flow problem as

$$E' \equiv \begin{cases} \left(\dfrac{L_2}{L_1}\right)^2 \dfrac{\partial}{\partial\bar{x}}\left(\beta\dfrac{H}{H_r}\dfrac{\partial\bar{U}}{\partial\bar{x}}\right) + \dfrac{\partial}{\partial\bar{y}}\left(\beta\dfrac{H}{H_r}\dfrac{\partial\bar{U}}{\partial\bar{y}}\right) \approx 0 \\[2mm] 0 \le \bar{x} \le 1, \quad 0 \le \bar{y} \le 1 \\[1mm] \text{Boundary conditions of } \bar{U} \end{cases} \tag{9-34}$$

Here

$$\bar{x} = \frac{x}{L_1}, \qquad \bar{y} = \frac{y}{L_2}$$

with L_1 and L_2 being maximum dimensions of the flow region, and

$$\bar{U} = \frac{U - U_1}{U_2 - U_1} \tag{9-35}$$

where U_1 is the smallest value of U in the field and U_2 is the largest value. Similarly a dimensionless system for the electrical problem is

$$G' \equiv \begin{cases} \left(\frac{l_2}{l_1}\right)^2 \frac{\partial}{\partial \bar{x}} \left(\frac{\sigma h}{\sigma_r h_r} \frac{\partial \bar{V}}{\partial \bar{x}}\right) + \frac{\partial}{\partial \bar{y}} \left(\frac{\sigma h}{\sigma_r h_q} \frac{\partial \bar{V}}{\partial \bar{y}}\right) \approx 0 \\ 0 \leq \bar{x} \leq 1, \qquad 0 \leq \bar{y} \leq 1 \\ \text{Boundary conditions on } \bar{V} \end{cases} \tag{9-36}$$

Here the definitions of $\bar{x}$, $\bar{y}$, l_1, l_2, $\bar{V}$, V_1 and V_2 are exactly of the same form as in E' above.

Thus, if the two systems have similar geometry and boundary conditions, we see that

$$\bar{U}(\bar{x}, \bar{y}) = \bar{V}(\bar{x}, \bar{y}) \tag{9-37}$$

if

$$\frac{\beta H}{H_r}(\bar{x}, \bar{y}) = \frac{\sigma h}{\sigma_r h_r}(\bar{x}, \bar{y}) \tag{9-38}$$

That is, $\beta H / H_r$ and $\sigma h / \sigma_r h_r$ are the same functions of the dimensionless coordinates. Here σ_r and h_r are reference values, the values at $x = 0$, $y = 0$, for example.

This analysis shows how an electrical system can be used to represent a fluid-flow system.

Usually the fluid-flow problem is one in which both K and H are variable. However, in the electrical analog only variations in h need be included; thus $\sigma = \sigma_r$ everywhere. Most often the electrical analog takes the form of a tank made of insulating material filled with a solution of sodium chloride in water. Variations in depth then represent variations in KH for the fluid system.

If large variations in K exist in the flow system (discontinuities in K, for example) then large variations in h must be introduced. This would introduce large vertical components of flow in the electrical system. To eliminate this, one can place many small stiff wires standing vertically in the tank at the discontinuity. Two rows of such wires, one on the low side and the other on the high side of the discontinuity forces the lines of flow to lie in a horizontal plane. Such wires can be distributed uniformly over the tank to maintain approximately horizontal flow everywhere, at least to a sufficient degree for most applications.

Though the formulation of the analog given here yields $\bar{U} = \bar{V}$, it still remains to consider the relation between flow rate per unit area in the fluid system and the electrical current density.

From equations (9-29) and (9-31) and the definitions of the dimensionless variables we have

$$\hat{\bar{m}} = \left(\frac{H_r}{H} \frac{U_2 - U_1}{L_1}\right)\left(\frac{l_1}{V_2 - V_1}\right)\left(\frac{h}{h_r \sigma_r}\right)\hat{j} \tag{9-39}$$

as the relation between $\hat{\bar{m}}$ and $\hat{j}$.

In such analog models the sources and sinks are metal electrodes connected to an alternating current power supply; usually sixty cycle current is used.

The particular analog model described here is most useful in the study of waterflooding problems for unit mobility ratio. Thus since

$$\frac{\partial \bar{U}}{\partial \bar{x}} = \frac{\partial \bar{V}}{\partial \bar{x}}, \quad \frac{\partial \bar{U}}{\partial \bar{y}} = \frac{\partial \bar{V}}{\partial \bar{y}} \tag{9-40}$$

measurements of $\nabla \bar{V}$ by potential probes can be used to plot the development of a flood front. Several other types of analogs have been developed for this and similar flow problems.[5, 9, 10]

9.60: Example of Scaling When the Equations for the System are not Known

To illustrate some aspects of the scaling problem encountered when the equations describing the physical system are not known, the following problem is considered.*

A model is to be constructed to study the slow, steady flow of a viscous, incompressible fluid through a capillary tube of axial symmetry whose radius is a sinusoidal function of distance along the axis. The tube is to be placed horizontally so that gravity effects can be neglected. Since the flow is to be slow, it is reasonable to assume that inertial effects can also be neglected.

The parameters characterizing the geometry are the minimum diameter, d_1, the maximum diameter, d_2, the distance between successive maxima in diameter, l, and the length of the tube, H. The variable to be studied is flow rate through the tube, Q, as a function of the pressure difference

---

* Actually the Navier-Stokes equations could be used for similarity analysis.

across the tube, Δp. Since the fluid is viscous, the viscosity of the fluid, μ, is an important parameter.

By using knowledge of flow through a straight capillary as an intuitive guide, it can be seen that the above quantities seem to be the only variables and parameters pertinent to the problem. Thus, following the outline given in section 9.20 on general theory, one takes

$$Q = \text{dependent variable}$$

$$\Delta p = \text{independent variable} \tag{9-41}$$

$$d_1, d_2, l, H, \text{ and } \mu \text{ are parameters}$$

Here there are seven quantities and three independent physical dimensions. The basic dimensions are taken as force, length, and time, and are denoted by F, L, and T. According to the π-Theorem there must exist four independent dimensionless groups for the problem. These groups are denoted by π_i, $i = 1, 2, 3, 4$. It is noted that the dimensions of the variables and parameters are

$$[Q] = \frac{L^3}{T}$$

$$[\Delta p] = \frac{F}{L^2}$$

$$[d_1] = [d_2] = [l] = [H] = L \tag{9-42}$$

$$[\mu] = \frac{FT}{L^2}$$

Here [ ] means "dimensions of," and F, L and T stand for force, length and time, respectively.

Following the procedure outlined in the Appendix for applying the π-Theorem would lead to the necessity of solving many simultaneous equations to construct the π_i. The procedure is greatly simplified by noting that from equations (9-42)

$$\pi_1 = \frac{d_1}{d_2}$$

$$\pi_2 = \frac{d_1}{l} \tag{9-43}$$

$$\pi_3 = \frac{H}{l}$$

are obvious as acceptable dimensionless groups. This leaves one group to be determined.

Obviously, both Q and Δp must be included in this last group. Also, since L does not have the same power in these two variables, one of the parameters of dimension L must be included. Further, since T occurs in Q only, and F in Δp only, μ must be included. Thus, one takes as π_4

$$\pi_4 = Q^{\gamma_1}(\Delta p)^{\gamma_2}\, d_1^{\gamma_3}\mu^{\gamma_4} \tag{9-44}$$

where γ_1, γ_2, $\cdots$ are numerical exponents. This can be written in terms of dimensions as

$$[\pi_4] = 0 = \left[\frac{L^3}{T}\right]^{\gamma_1}\left[\frac{F}{L^2}\right]^{\gamma_2}[L]^{\gamma_3}\left[\frac{FT}{L^2}\right]^{\gamma_4} \tag{9-45}$$

which yields

$$\left.\begin{aligned}
3\gamma_1 - 2\gamma_2 + \gamma_3 - 2\gamma_4 &= 0 \quad \text{(L dimension)} \\
-\gamma_1 + \gamma_4 &= 0 \quad \text{(T dimension)} \\
\gamma_2 + \gamma_4 &= 0 \quad \text{(F dimension)}
\end{aligned}\right\} \tag{9-46}$$

Since (as is usually the case) there are four unknowns and only three equations, one of the γ's can be selected at will. Since it is Q that is of most interest, take $\gamma_1 = 1$, and from equations (9-46)

$$\gamma_4 = 1, \gamma_2 = -1, \gamma_3 = -3 \tag{9-47}$$

Thus, have

$$\pi_4 = \frac{Q\mu}{d_1^3\Delta p} \tag{9-48}$$

as the fourth dimensionless group.

Always when dealing with dimensionless groups any one of the groups can be replaced by an algebraic combination of the others. Thus, in lieu of π_4 above, take as the modified π_4

$$\pi_4 = \left(\frac{Q\mu}{d_1^3\Delta p}\right)\frac{\pi_3}{\pi_2} = \frac{Q\mu H}{d_1^4\Delta p} \tag{9-49}$$

Thus, it is concluded that the slow, steady flow of a viscous, incompressible fluid through a tube of axial symmetry and sinusoidally varying diameter must obey an equation of the form

$$\frac{Q\mu H}{d_1^4\Delta p} = f\left(\frac{d_1}{d_2}, \frac{d_1}{l}, \frac{H}{l}\right) \tag{9-50}$$

where f is a dimensionless function of the three indicated parameters.

Not only has this analysis yielded scaling groups for the problem, but it has also indicated the form of the solution of the problem. In fact, this can be carried even further. If $d_1 = d_2$, then this problem becomes just that of a straight capillary tube and the mathematical solution of this problem is known.

$$\left.\begin{array}{l} \text{for } d_1 = d_2 \\[2mm] \dfrac{Q\mu H}{d_1{}^4\Delta p} = \dfrac{\pi}{128} \end{array}\right\} \tag{9-51}$$

Thus, equation (9-51) can be written in the form

$$\frac{Q\mu H}{d_1{}^4\Delta p} = \frac{\pi}{128} f'\left(\frac{d_1}{d_2}, \frac{d_1}{l}, \frac{H}{l}\right) \tag{9-52}$$

where now the unknown function, f', approaches unity as d_1 approaches d_2. It can be argued further on the basis of analogy that when H is much greater than l the function f' should be relatively insensitive to the value of H/l.

$$\left.\begin{array}{l} \text{for } H \gg l \\[2mm] \dfrac{Q\mu H}{d_1{}^4\Delta p} \approx \dfrac{\pi}{128} f''\left(\dfrac{d_1}{d_2}, \dfrac{d_1}{l}\right) \end{array}\right\} \tag{9-53}$$

where two primes are used to distinguish this function from f'.

To summarize the analysis: If model and prototype have the same shape, and d_1/d_2, d_1/l and H/l have the same values in the model and the prototype, then $Q\mu H/d_1{}^4\Delta p$ will have the same value in model and prototype. Furthermore, the data obtained on the model can be correlated in the form given by equation (9-52), where the function f' approaches unity as $d_1 \to d_2$ and also f' is relatively independent of H/l when H/l is large.

It must be noted here that the neglect of inertial forces in both model and prototype must be justified by experiment or some theoretical means. This is discussed in the next section.

This example should illustrate the power and utility of this type of analysis.

9.70: Partial Similarity in Scaled Models

In the design of laboratory models of size different from the prototype system, it is ordinarily desirable to maintain exact dynamic similarity, if possible. This means that all dimensionless parameters are maintained at values equal to those of the prototype. If this can be done, the model measurements yield numbers for the dimensionless variables equal to those which would occur in the prototype. Most frequently this cannot be accomplished for at least one of two reasons: (1) the nature of the prototype system is not sufficiently well understood to permit confident formulation and subsequent translation into a model, (2) even when the problem is clearly defined, it may be impossible to maintain full dynamic similarity with a change in size.

The first of these difficulties often arises in exploratory research. It sometimes requires a keen physical insight and good luck to set up a laboratory model which displays dominant behavior like that of a prototype.

The same talents are useful when it is desired to devise a simplified model of a well-defined but excessively complex physical system. Here the researcher must discover which of the many known interacting effects may be neglected and which are essential for representing the dominant behavior of the system. In deducing this, it may be that the simplifications achieved are sufficient even to permit a mathematical solution of the problem.

Most of the points mentioned above can be illustrated with the aid of the problem discussed in the last section, flow through a capillary with sinusoidal variations in cross section.

First, consider the first type of difficulty discussed above, namely insufficient knowledge of the system. In the scaling analysis for this problem, inertial effects were completely neglected; consequently, the scaled model formulated there has only partial similarity to a prototype, because the inertial properties of model and prototype may be radically different. Since the question to be studied was the dependence of Q on Δp, this lack of similarity is not a handicap provided the dependence of Q on Δp is independent of inertial effects in the range of values of Q of interest. Thus, some sort of criterion is needed to determine the importance of inertia in this relationship. This is where some basic physics may be applied.

The inertial properties of the flow can obviously be characterized by the kinetic energy of the fluid; thus, a point of attack on the problem would be the application of the principle of the conservation of energy. Since no potential energy is involved, this can be stated as

$$\begin{pmatrix} \text{work done on} \\ \text{fluid column} \\ \text{of length } H \end{pmatrix} = \begin{pmatrix} \text{increase of} \\ \text{kinetic energy} \\ \text{of fluid column} \\ \text{of length } H \end{pmatrix} - \begin{pmatrix} \text{energy loss} \\ \text{due to friction} \\ \text{in length } H \end{pmatrix} \qquad (9\text{-}54)$$

If the density of the fluid is denoted by ρ, then

$$\begin{pmatrix} \text{kinetic} \\ \text{energy} \end{pmatrix} \approx (\text{mass})(\text{velocity})^2 \propto (\rho d_1{}^2 H)\left(\frac{Q}{d_1{}^2}\right)^2$$

$$(\text{work done}) \approx (\text{force})(\text{distance}) \propto (\Delta p d_1{}^2)(H)$$

$$(\text{energy loss}) \approx (\text{viscosity})(\text{cross-section area})\begin{pmatrix} \text{radial} \\ \text{velocity} \\ \text{gradient} \end{pmatrix}(\text{distance}) \qquad (9\text{-}55)$$

or,

$$(\text{energy loss}) \propto \mu d_1{}^2\left(\frac{Q}{d_1{}^3}\right)(H)$$

In all of these relations, ∝ means "proportional" to. Furthermore, if these proportionalities are to be replaced by equalities, the constants of proportionality must be, because of the dimensions in these relations, dimensionless functions of d_1/d_2, d_1/l, and H/l.

Thus

$$\Delta p d_1{}^2 H = f_1 \left(\frac{d_1}{d_2}, \frac{d_1}{l}, \frac{H}{l} \right) \left(\frac{Q \mu H}{d_1} \right) + f_2 \left(\frac{d_1}{d_2}, \frac{d_1}{l}, \frac{H}{l} \right) \left(\frac{\rho Q^2 H}{d_1{}^2} \right) \tag{9-56}$$

or, with obvious rearrangement and new functions of proportionality

$$\frac{Q \mu H}{d_1{}^4 \Delta p} = \frac{\pi}{128} f' \left(\frac{d_1}{d_2}, \frac{d_1}{l}, \frac{H}{l} \right) \left[1 - \left(\frac{\rho Q^2}{d_1{}^4 \Delta p} \right) F \left(\frac{d_1}{d_2}, \frac{d_1}{l}, \frac{H}{l} \right) \right] \tag{9-57}$$

Thus, the requirement of conservation of energy has led to a more complete formulation of the problem. In particular, inertial effects are now included. Equation (9-57) shows how the model data should be correlated when ρ is included in the problem.

The desired criterion as to the importance of inertia can thus be formulated as

$$\left(\begin{matrix} \text{kinetic} \\ \text{energy} \end{matrix} \right) \ll \left(\begin{matrix} \text{energy loss} \\ \text{due to viscosity} \end{matrix} \right) \tag{9-58}$$

or approximately

$$(\rho d_1{}^2 H) \left(\frac{Q}{d_1{}^2} \right)^2 \ll \mu d_1{}^2 \left(\frac{Q}{d_1{}^3} \right) H \tag{9-59}$$

which yields

$$\frac{\rho Q}{\mu d_1} \ll 1 \tag{9-60}$$

where consistent units must be employed. When equation (9-60) is satisfied the formulation given previously, neglecting inertia, is satisfactory and the model with partial similarity adequately represents the prototype. It should be noted that experimental data may show that this restriction can be relaxed to some extent.

The second kind of difficulty might arise if, for example, the flow through a sinusoidal capillary 1,000 feet long were to be studied with a model. Obviously, if a model only a few feet long were employed, the diameter of the model capillary would have to be vanishingly small in order to maintain $(H/l)/(d_1/l)$ at the same value, and this would be impractical, if not impossible. Here, physical intuition can be applied to reduce the dilemma. Analogy to a straight capillary suggests that the flow behavior should be relatively insensitive to the value of H/l for large values of H/l, thus a few trial experiments with models having different values of H/l should show that for some value of H/l, say $(H/l)_c$, any further increase in H/l does not alter the value of $Q\mu H/d_1^2 \Delta p$. Then, a model similar in all respects but the value of H/l to the prototype could be used, and this model with partial similarity would correctly predict the dependence of Q on Δp.

These examples illustrate some of the variety of techniques required in the design of models having partial similarity.

The example employed here and in the preceding section has applications to the study of flow through porous materials. A flow channel in a porous medium has variations in cross section along its length. Thus to apply the above results to flow through a porous medium, d_1 could correspond to the average pore diameter, d_2 to the maximum pore diameter, l to the standard deviation of the pore-size distribution and H to τL, where τ is tortuosity and L the sample length.

9.80: Special Aspects of Model Scaling for Porous Media

Certain special features of model scaling for studies of fluid flow in porous media should be noted. The technique for handling relative permeabilities and capillary pressures has already been noted in Section 9.30. However the case of anisotropic media was not considered.

It was pointed out in section 3.70 that a particular transformation of coordinates put the equations of flow for a homogeneous fluid through an anisotropic medium into the form corresponding to an isotropic medium. Also it was pointed out in section 3.32 that the assumption that relative permeability is independent of direction in anistoropic media is reasonable.

Thus for multiphase flow in anisotropic media a choice of dimensionless coordinates of the form.

$$\bar{x}_i = \frac{x_i}{L_i}, \qquad i = 1, 2, 3 \tag{9-61}$$

will introduce the groups

$$\frac{K_1 L_2^2}{K_2 L_1^2} \quad \text{and} \quad \frac{K_1 L_3^2}{K_3 L_1^2} \tag{9-62}$$

into the scaling laws. This will apply if relative permeabilities are assumed isotropic. Consequently, an isotropic model can represent an anisotropic prototype. For example

$$\frac{L_2 m}{L_1 m} = \frac{K_{1p} L_{2p}^2}{K_{2p} L_{1p}^2} \tag{9-63}$$

where m refers to model and p to prototype. That is

$$K_{2m} = K_{1m} \qquad (9\text{-}64)$$

and the model is isotropic.

Another special feature of model scaling as applied to flow through porous media arises as a result of the complementary microscopic and macroscopic descriptions of flow in such media. For most flow problems only the macroscopic description need be considered but in some cases the microscopic features must also be considered.

Thus in the study of miscible displacement the dispersion factor (section 8.30) is determined by pore size and average pore velocity. But the macroscopic flow geometry is determined by the macrosocpic distribution of permeability and the boundary conditions. Generally, dispersion will be an important factor in small scaled models but will not be significant for systems the size of a petroleum reservoir Thus, one might formulate the description applicable to a reservoir and deduce scaling laws from this, which would neglect dispersion. But a laboratory model would exhibit excessive dispersion and hence not represent the prototype unless dispersion were considered. Thus the dispersion coefficient must be made very small in the model in order to represent the reservoir.

Flow problems involving both liquids and gases in porous media present rather severe limitations on model design. This is so because of the particular pressure-volume relations applicable to gases. As a result, such models must generally be constructed to operate at the same pressure level as the prototype system.

Due to the requirement of continuity of fluid pressures, and hence capillary pressure, across discontinuities in permeability the presence of such features in flow systems always introduces additional scaling groups. Also the requirement of continuity of flow normal to such discontinuities introduces additional scaling groups. Consequently, one must be careful to include all boundary conditions as well as all differential equations in deriving scaling requirements by similarity techniques.

9.90: Results of Model Studies of Flow Through Porous Media

Numerous model studies of various problems involving the flow of fluids through porous materials have appeared in the literature, far too many to summarize in a text of this nature. Therefore, the reader is referred to the references at the end of this chapter for such detailed results.[2, 3, 4, 5, 7, 8, 9, 10]

APPENDIX TO CHAPTER 9

The Buckingham π-Theorem

This theorem and many of its implications are discussed at great length in reference.[1] Here a statement of the theorem is given.

"If the equation

$$f(a_1, a_2, a_3, \cdots, a_n) = 0$$

is a complete equation, and also the only equation relating the n quantities $a_1, a_2, \cdots$, *which have a total of D independent dimensions; then the equation*

$$F(\pi_1, \pi_2, \cdots \pi_{n-D}) = 0$$

where the π_i are $n - D$ dimensionless products of the a_1, a_2, etc., is completely equivalent to the given equation."

To apply this theorem, let $d_1, d_2, \cdots, d_D$ be the independent dimensions in the equation $f = 0$. In terms of these, let the dimensions of a_j be represented as

$$[a_j] = d_1^{x_{1j}} d_2^{x_{2j}} d_3^{x_{3j}} \cdots d_D^{x_{Dj}}$$

where the x_{ij} are numerical exponents.

Then, the dimensions of π_k are

$$[\pi_k] = [a_1]^{y_{1k}} [a_2]^{y_{2k}} \cdots [a_n]^{y_{nk}}$$

Since π_k must be dimensionless, it follows that

$$x_{11} y_{1k} + x_{12} y_{2k} + \cdots + x_{1n} y_{nk} = 0$$
$$x_{21} y_{1k} + x_{22} y_{2k} + \cdots + x_{2n} y_{nk} = 0$$
$$\vdots$$
$$x_{D1} y_{1k} + x_{D2} y_{2k} + \cdots + x_{Dn} y_{nk} = 0.$$

There are $n - D$ such sets of equations, each such set to be solved for the y's for each π_k. In general, several of the y's for each π_k must be set equal to zero in order for the π_k's to be independent. The reader is referred to the literature for the details of this type of analysis.

EXERCISES

1. Use the principle of similarity with p_c defined by equation (9-19) as a function of $\bar{S}_w'$ and k_o defined by equation (9-18) to show that the rate of oil flow from a linear system of length L by countercurrent imbibition is proportional to

$$\gamma \cos \theta \sqrt{\frac{K_{ro}^2}{K}} \bigg/ L$$

(See equations 6-96 and 6–97 in Chapter 6.)

2. Show that suitable scaling groups for two-dimensional miscible displacement in an isotropic porous medium are

$$\sqrt{\frac{K}{\phi}} \frac{q}{\phi L^2 D_1'}, \quad \frac{\mu_2}{\mu_1}, \quad \frac{(\rho_2 - \rho_1)gKL}{q\mu_1}, \quad \frac{H^2}{K}, \quad \frac{D_2'}{D_1'}, \quad \frac{H}{L}$$

Here L is length of sample, H is height of sample, q is injection rate and K is permeability. ρ_2, ρ_1, μ_2, μ_1, D_2', and D_1' are the density, viscosity and diffusion constant for the respective fluids and ϕ is the porosity of the sample. Present arguments to show that with these scaling requirements satisfied the concentration is the same function of x/L, y/L and $D_1't/L^2$ in two systems. Would fingering and segregation effects be scaled as well as dispersion effects?

References

1. Bridgman, P. W., "Dimensional Analysis," Yale Univ. Press, New Haven (1931).
2. Collins, R. E., and Perkins, F. M., To be published, *J. Petrol. Tech.* Sept. 1960.
3. Craig, F. F., Jr., Sanderlin, J. L., Moore, D. W., and Geffen, T. M., *Trans. AIME*, **210**, 275 (1957).
4. Gaucher, D. H., and Lindley, D. C., To be published, *J. Petrol. Tech.* (1960).
5. Landrum, B. L., Flanagan, D. A., Norwood, B. D., and Crawford, P. B., *Trans. AIME*, **216**, 33 (1959).
6. Leverett, M. C., *Trans. AIME*, **142**, 152 (1941).
7. Matthews, C. S., and Fischer, M. J., *Trans. AIME*, **207**, 111 (1956).
8. Matthews, C. S., and Lefkovits, H. C., *Trans. AIME*, **207**, 265 (1956).
9. Wyckoff, R. D., and Botset, H. G., *Physics*, **5**, 265 (1934).
10. Wyckoff, R. D., and Reed, D. W., *Physics*, **6**, 395 (1935).

10. FLOW WITH CHANGE OF PHASE

10.10: Flow with Change of Phase; Solution-Gas Drive

No text on the flow of fluids through porous materials would be complete without at least some discussion of the primary mechanism of petroleum production called solution-gas drive.

In previous chapters various flow regimes have been considered, however, none of these involved a change of phase; gas remained gas and liquid remained liquid. Here phase transitions in the flowing fluids will be considered with particular consideration of what is usually called solution-gas drive. First, it is necessary to consider phase equilibrium and phase transitions in stationary fluid systems.

It is not the intent of this chapter to present an exhaustive treatment of all aspects of solution-gas drive. For example, phase transitions as occur in surface separators are not discussed here. The intent is to make clear the processes which occur within the porous medium and to clarify the nature of various approximations often employed in the study of this process.

10.20: Phase Equilibrium*

Partition Factors and Equilibrium Ratios. By way of introduction to more complex systems and to establish some basic concepts of importance in all considerations of phase equilibrium the equilibrium of a single component system will be considered.

A single component system can exist in three possible phases; solid, liquid or gas. The state of a single component system is described by an equation of state, pressure p as a function of volume V and temperature T, for example. However, the equation of state is different for each phase of the system. At certain values, or over certain ranges of values, of the thermodynamic variables two, or even all three, phases can exist simultaneously. Let

$$F_V(p, V_V, T) = 0 \tag{10-1}$$

be the equation of state for the vapor phase and

$$F_L(p, V_L, T) = 0 \tag{10-2}$$

be the equation of state for the liquid phase.

---

* For a more detailed exposition, see reference 1.

If both liquid and vapor exist in equilibrium with each other, then P and T are no longer independent variables. This is the vapor-pressure curve. Such reductions of the number of independent variables are forecast in general by Gibb's phase rule.[2]

$$F = c - p + 2$$

Here F is the number of degrees of freedom (independent variables), c is the number of components in the system and p is the number of phases existing in equilibrium. Thus, for a single component all three phases can exist in equilibrium at only one pressure, temperature and volume. This is the "triple point."

A typical pressure-temperature diagram for a single-component system is shown in Figure 10-1. The three curves shown here divide the plane into three regions, each region being a single-phase region. Each curve gives the values of p and T at which the two adjacent phases can exist in equilibrium, and the intersection of these curves is the triple point.

In flow problems we are concerned only with liquids and vapors, or gases. In this case it is frequently convenient to represent the system on a pressure-volume diagram. Such a diagram is shown in Figure 10-2.

Here two isotherms are shown. For temperature T_1 the curve is entirely in the vapor region. For temperature T_2 the isotherm is in the vapor region for large values of V. As V is reduced a point on this curve moves to a point on the dew-point curve. Here liquid starts to form. After this curve is crossed liquid and vapor exist in equilibrium. As the bubble-point curve is passed all the vapor condenses and the whole system is in the liquid state. The point c is the critical point. A unique characteristic of a single-

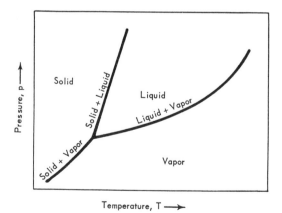

Figure 10–1. Pressure-temperature diagram for a single-component system.

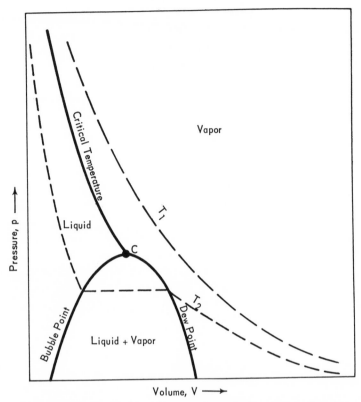

Figure 10–2. Pressure-volume diagram for a single-component system.

component system is that at a given temperature the bubble-point pressure, dew-point pressure and vapor-pressure are all equal.

Now a given mass of a single-component system may exist partly as liquid and partly as vapor. In Figure 10-1, this is shown on the vapor-pressure curve. Thus if another component is added, which is soluble in both liquid and vapor of the first component, we might ask what is the relative distribution of the second component between the two phases. For example, consider water in equilibrium with its vapor and the addition of carbon-dioxide gas to the system. The CO_2 may partially go into solution in the liquid and partially remain in the gaseous state with the water vapor.

To answer this question of relative distribution consider the more general problem of a substance being dissolved in two other substances. Say, some of a substance A is dissolved in each of two substances, B and C.

Define the solubility of A in B as

$$s_{AB} = \frac{n_A}{n_A + n_B} \tag{10-3}$$

where n_A is number of moles of A dissolved in n_B moles of B. Similarly

$$s_{AC} = \frac{n_A'}{n_A' + n_C} \tag{10-4}$$

where n_A' moles of A are dissolved in n_C moles of C. These solubilities will be functions of pressure and temperature, and possibly other factors also.

The ratio

$$E_{B,C}^A = \frac{s_{AB}}{s_{AC}} \tag{10-5}$$

is the partition factor for A between B and C.

Now if substance A is a fluid (liquid, vapor, or both), B is a liquid and C is its vapor then all of A is either in B or C; i.e. either in liquid or vapor phase. In this case, we call

$$E_A = \frac{C_{AV}}{C_{AL}} \tag{10-6}$$

the "equilibrium ratio" (K is usually used for this quantity but to avoid confusion with permeability we use E). Here C_{AV} is the mole fraction of the vapor which is A and C_{AL} is the mole fraction of the liquid which is A.

For each additional component an equilibrium ratio can be defined in the same way. Thus for the ith component of an N-component system

$$E_i = \frac{C_{iV}}{C_{iL}} \tag{10-7}$$

where

$$C_{iV} = \frac{n_{iV}}{n_{1V} + n_{2V} + \cdots + n_{NV}} \tag{10-8}$$

and

$$C_{iL} = \frac{n_{iL}}{n_{1L} + n_{2L} + \cdots + n_{NL}} \tag{10-9}$$

These equilibrium ratios are functions of pressure and temperature, and also of the over-all composition of the system. This over-all composition is specified by giving the mole fractions of the whole system for each component. Thus

$$C_i = \frac{n_i}{n_1 + n_2 + \cdots + n_N}, \qquad i = 1, 2, \cdots N \tag{10-10}$$

where n_i is the number of moles of the ith component in the whole system. i.e.

$$n_i = n_{iV} + n_{iL} \tag{10-11}$$

If the vapor obeys the ideal gas law and Dalton's law of partial pressures, and the liquid (and its components) obey Raoult's solution law, then these equilibrium ratios can be theoretically computed. In practice, however, this is not the case and the equilibrium ratios must be empirically determined.

Now defining C_V as the mole fraction of the composite system in the vapor state and C_L as the mole fraction of the composite system in the vapor state, we have

$$C_V + C_L = 1 \tag{10-12}$$

Also

$$C_i = C_{iL}C_L + C_{iV}C_V \tag{10-13}$$

Thus using equation (10-7) to eliminate C_{iV} here

$$C_{iL} = \frac{C_i}{1 + C_V(E_i - 1)} \tag{10-14}$$

Then since

$$\sum_{i=1}^{N} C_{iL} = 1 \tag{10-15}$$

it follows that

$$\sum_{i=1}^{N} \frac{C_i}{1 + C_V(E_i - 1)} = 1 \tag{10-16}$$

Similarly

$$C_{iV} = \frac{C_i E_i}{1 + C_V(E_i - 1)} \tag{10-17}$$

and

$$\sum_{i=1}^{N} \frac{C_i E_i}{1 + C_V(E_i - 1)} = 1 \tag{10-18}$$

Dew-Point and Bubble-Point of Multicomponent System. Now just as for a single-component system the dew-point is defined as that state in which the system is entirely in the vapor state and any slight increase in pressure (or reduction in volume) produces a liquid phase at constant temperature. Similarly, at fixed pressure and volume a slight reduction in temperature will produce a liquid phase.

Also as for a single-component system the bubble-point is defined as that state in which the system is entirely in the liquid state and any slight reduction in pressure (or increase in volume) at fixed temperature produces a vapor phase. Similarly, at fixed pressure and volume a slight increase in temperature produces a vapor phase.

At the *dew-point* $C_v = 1$ and equation (10-16) gives

$$\sum_{i=1}^{N} \frac{C_i}{E_i} = 1 \qquad (10\text{-}19)$$

Similarly at the *bubble-point* $C_V = 0$ and equation (10-18) gives

$$\sum_{i=1}^{N} C_i E_i = 1 \qquad (10\text{-}20)$$

and, in both cases, we also have

$$\sum_{i=1}^{N} C_i = 1 \qquad (10\text{-}21)$$

Phase Distribution of Multicomponent Systems. In the section on partition factors and equilibrium ratios the question of the distribution of a component between vapor and liquid phases was considered. Often a question of equal importance is how much of a multicomponent system exists in each phase at a given pressure and temperature. This question can be investigated by means of the relations involving equilibrium ratios but a more direct empirical approach is possible. Before considering this question, it is important to emphasize that under some circumstances only one phase exists, just as for a single-component system. This is indicated in Figure 10-3.

Referring to this diagram consider the isotherm, T_1. As the pressure is reduced, at C only liquid exists, at B both liquid and vapor exist and ually at A only vapor exists. On the isotherm T_2 the system is entirely in the vapor phase at C', as the pressure is reduced the two-phase region is entered and at B' both phases exist. This formation of liquid on reduction of pressure is called *retrograde condensation*. As the pressure is further reduced along T_2 the system crosses the dew-point curve and at A' only vapor exists.

Considering pressure reduction along an isotherm observe that for the isotherm T_3 the pure vapor region is not reached even at $p = 0$. This is typical of most petroleum crude oils. For gas and gas-condensate reservoirs on the other hand the dew-point curve is well above $p = 0$ for all temperatures of interest.

At any stage of pressure reduction the liquid phase could be isolated om the free vapor phase. If the pressure on this liquid is reduced to atmospheric pressure and the temperature brought to standard atmospheric temperature some vapor would be evolved. If the vapor is continuously removed from contact with the remaining liquid, as it is formed, the process

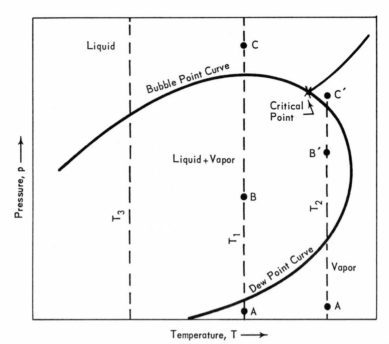

Figure 10–3. Pressure-temperature diagram for a multicomponent system showing the phase-boundary curves.

is called *differential vaporization*. If the evolved vapor is not removed continuously, it is a *flash vaporization*. In either case when standard atmospheric conditions are reached a unique volume of vapor will result, and also a unique volume of residual liquid. We consider this vapor, or gas, as having been dissolved in this volume of liquid at the original pressure and temperature. Calling the residual liquid, oil, with volume V_o and the vapor, gas, with volume V_g (both measured at atmospheric conditions), define the *solubility* as

$$s(p, T) = \frac{V_g \text{ (standard conditions)}}{V_o \text{ (standard conditions)}} \tag{10-22}$$

Also noting that the volume of liquid resulting from this solution of gas in oil is not V_o, we define the formation volume factor for oil as

$$\beta_o(p, T) = \frac{V_L(p, T)}{V_o \text{ (standard conditions)}} \tag{10-23}$$

where $V_L(p, T)$ denotes liquid volume at p and T.

Since for differential and flash vaporizations the liquid is not exposed

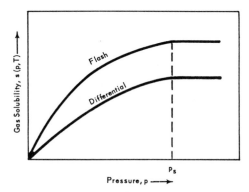

Figure 10–4. Gas solubility versus pressure for a typical petroleum hydrocarbon system. p_s is the saturation pressure. The two curves represent differential and flash-equilibrium conditions.

to the same composition of gas at any stage, s and β_o are not the same functions of p and T for the two processes. (Also the equilibrium ratios, E_i, defined under partition factors, are not the same for both processes.)

Typical plots of s versus p for fixed T are shown in Figure 10–4 for the two processes. Generally, solubility decreases with increase in the density of the oil.* For a system of fixed composition a pressure is reached at which no more gas will go into solution. This is the saturation or bubble point pressure and is indicated as p_s in Figure 10–4. For $p > p_s$, s is constant. The oil is then saturated.

The dependence of β_o on pressure is indicated in Figure 10–5. Here again a distinction between flash and differential vaporization is necessary. Below the saturation pressure the liquid increases in volume as pressure increases and more gas goes into solution. Above the saturation pressure no additional gas goes into solution and simple compression of liquid occurs to yield a decrease in β_o.

In addition to these characteristics of the phase distribution in a multi-component system, it is convenient to define another property. Consider the vapor which exists in equilibrium with liquid at any temperature and pressure. If this vapor (or gas) is isolated from the liquid and brought to atmospheric conditions its volume will change. Thus define the formation volume factor for gas as

$$\beta_g = \frac{V_g(p, T)}{V_g \text{ (standard conditions)}} \tag{10-24}$$

* In the petroleum industry, oil density is expressed as "°API gravity," defined as:

$$°\text{API gravity} = \frac{141.5}{\text{specific gravity } 60° \text{ F}} - 131.5$$

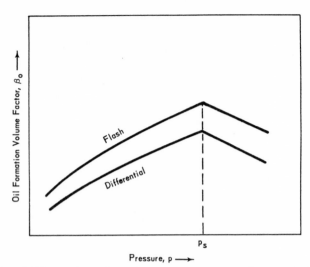

Figure 10–5. Oil formation volume factor versus pressure for a typical petroleum hydrocarbon system. p_s is the saturation pressure. The two curves represent differential and flash-equilibrium conditions.

Note that β_g depends on p and T not only because of the expansibility of the gas but also because the physical composition of the gas is different at different p and T.

10.30: Supersaturation and Undersaturation

The relations between gas and liquid compositions discussed in section 10.20 are only valid for equilibrium conditions. That is

$$E_i = C_{iV}/C_{iL}$$

was stated to be a constant for given pressure and temperature. This is true only when the system has been maintained at fixed temperature and pressure for a sufficient period of time.

If a system of several components is in equilibrium at a pressure p_1 and temperature T and the pressure is suddenly changed to a different value, p_2, the following sequence of events occurs, provided $p_2 - p_1$ is small.

In a small region in the neighborhood of the gas-liquid interface the equilibrium distribution of components between gas and liquid phases corresponding to the new pressure, p_2, is instantaneously established. Thus a gradient of composition in each phase is established. This gives rise to diffusion within the individual phases. This diffusion persists until again a gas of uniform composition is in equilibrium with a liquid of uniform composition, and again the equations of section 10.20 are valid.

A similar process occurs if a gas of non-equilibrium composition is brought into contact with a liquid at fixed temperature and pressure. Diffusion begins and persists until equilibrium compositions of gas and liquid are attained.

The order of magnitude of time required for equilibrium composition to be attained by diffusion alone can be estimated as follows.

Suppose a very large quantity of a gas with a small excess of one component is brought into contact with a uniform layer of liquid having depth L. Diffusion of the excess component into the liquid will occur. Within the liquid this diffusion will be described by

$$D_i \frac{\partial^2 \bar{C}_{iL}}{\partial x^2} = \frac{\partial \bar{C}_{iL}}{\partial t}, \qquad 0 < x < L \tag{10-25}$$

where D_i is the diffusion coefficient, $x = 0$ is the gas-liquid interface and $\bar{C}_{iL}$ is the concentration of excess component in the liquid expressed as moles per unit volume.

If the equilibrium value of C_{iL} is small then the distinction between mole fraction and moles per unit volume is not significant. I.e.

$$C_{iL} = \frac{\bar{C}_{iL}}{\bar{C}_{iL} + \bar{C}_{xL}} \tag{10-26}$$

where C_{xL} is moles per unit volume of other components in the liquid. Thus

$$\bar{C}_{iL} = \bar{C}_{xL} \frac{C_{iL}}{1 - C_{iL}} \tag{10-27}$$

Thus, if the mole fraction of this component, C_{iL}, is always much less than one, we have

$$\bar{C}_{iL} \approx \frac{\rho_L}{M_L} C_{iL} \tag{10-28}$$

since $C_{xL} \approx \rho_L/M_L$, where ρ_L is liquid density and M_L is mean molecular weight of liquid. Since ρ_L/M_L is essentially constant, we can write

$$D_i \frac{\partial^2 C_{iL}}{\partial x^2} = \frac{\partial C_{iL}}{\partial t}, \qquad 0 < x < L \tag{10-29}$$

At the gas-liquid interface instantaneous equilibrium is attained

$$\frac{C_{iV}}{C_{iL}} = E_i \tag{10-30}$$

and if a very large quantity of gas is present and only a small quantity of oil, then C_{iV} remains essentially constant. Thus

$$C_{iL} = \frac{C_{iV}}{E_i} \approx \text{constant}, \quad \text{at} \quad x = 0 \tag{10-31}$$

The boundary conditions at $x = L$ are

$$\frac{\partial C_{iL}}{\partial x} = 0, \quad \text{at} \quad x = L \tag{10-32}$$

and at $t = 0$, we have

$$C_{iL} = 0, 0 < x < L, t = 0 \tag{10-33}$$

From the solution of this boundary-value problem, it can be shown that[3, 5]

$$\int_0^L C_{iL} \, dx \approx \frac{C_{iV}L}{E_i} \left(1 - \frac{8}{\pi^2} \exp -D_i t(\pi/2L)^2 \right) \tag{10-34}$$

approximately, or

$$Q_i = \frac{\rho_L C_{iV} A L}{M_L E_i} \left(1 - \frac{8}{\pi^2} \exp - D_i t(\pi/2L)^2 \right) \tag{10-35}$$

where Q_i is the total number of moles of the component i diffused through area A of gas-liquid interface in time t.

We can then define the saturation fraction, u_i, as the ratio of the quantity of component i in the liquid to the quantity in the liquid at equilibrium. Thus

$$u_i = 1 - \frac{8}{\pi^2} \exp - D_i t(\pi/2L)^2 \tag{10-36}$$

The liquid is said to be undersaturated when $u_i < 1$ and supersaturated when $u_i > 1$. This formulation could have been carried through considering a deficit of component i in the gas, or an excess in the liquid and the liquid would be supersaturated.

A typical value for D_i is about 2×10^{-8} sq ft/sec. Considering oil globules, or films, with L the order of pore dimensions in a porous rock, $L \approx 5 \times 10^{-4}$ ft we obtain

$$u_i \approx 1 - \frac{8}{\pi^2} e^{-0.2t} \tag{10-37}$$

Within about 10 to 20 seconds such an oil globule or film would be essentially completely saturated.

During two-phase flow in a porous medium a given element of gas will be in contact with a particular element of oil for a period of time roughly equal to the average transit time through a distance of one grain diameter, or one pore diameter. The relative velocity of the fluids is

$$| v_R | \approx \left| \frac{v}{\phi} f_v - \frac{v}{\phi} (1 - f_v) \right| \leq \frac{v}{\phi} \tag{10-38}$$

Thus, the contact time is

$$t \geq \frac{\phi \, d}{v} \tag{10-39}$$

where v is volume flow rate per unit area, ϕ is porosity, f_v is the fraction of gas in the flowing stream and d is pore diameter. For flow rates typical in petroleum reservoirs, $v \approx 1$ ft/day, and hence

$$t \geq 10 \text{ sec.} \tag{10-40}$$

Hence we can safely assume that for solution-gas drive in petroleum reservoirs local equilibrium between the gas and oil phases exists.

This is not true in laboratory flow experiments. Laboratory solution-gas drive experiments often give rise to flow rates of 50 to 100 ft/day and then the contact time in the pores is only of the order of $\frac{1}{10}$ sec which is not sufficient for equilibrium to be attained.

10.40: Formulation of the Solution-Gas Drive Process

The material presented in previous sections of this chapter can be used in conjunction with Darcy's law and general conservation principles to construct a general mathematical formulation of the solution-gas drive process.

During solution-gas drive there are two flowing fluids, a liquid and a gas. By Darcy's law the volume flux densities for these two fluids are

$$\hat{v}_g = -\frac{K_g}{\mu_g} (\nabla p_g + \hat{1}_v \rho_g g) \tag{10-41}$$

and

$$\hat{v}_L = -\frac{K_L}{\mu_L} (\nabla p_L + \hat{1}_v \rho_L g) \tag{10-42}$$

where L and g refer to liquid and gas respectively, and $\hat{1}_v$ is a unit vector directed vertically upward.

Here the pressures, p_g and p_L, are related by the capillary pressure

$$p_g - p_L = p_c \tag{10-43}$$

Also it is assumed that the fluid densities, ρ_g and ρ_L, and the fluid viscosities, μ_g and μ_L, are known functions of composition, pressure and temperature. Thus

$$\left.\begin{aligned}\rho_L &= \rho_L(C_{iL}\,,\,p_L\,,\,T)\\ \rho_g &= \rho_g(C_{ig}\,,\,p_g\,,\,T)\end{aligned}\right\} \tag{10-44}$$

and

$$\left.\begin{aligned}\mu_L &= \mu_L(C_{iL}\,,\,p_L\,,T)\\ \mu_g &= \mu_g(C_{ig}\,,\,p_g\,,\,T)\end{aligned}\right\} \tag{10-45}$$

where C_{iL} is mole fraction of liquid which is component i, and similarly for C_{ig}. ($i = 1, 2, \cdots N$.)

Now the total number of moles of a component, say the ith component, is the quantity which is conserved. The number of moles of i per unit bulk volume of porous medium is

$$\left(C_{iL}\,\frac{\rho_L}{M_L}\,S_L + C_{ig}\,\frac{\rho_g}{M_g}\,S_g\right)\phi \tag{10-46}$$

where ϕ is porosity, S_L and S_g are liquid and gas saturation, and M_L and M_g are mean molecular weights of liquid and gas, respectively. Thus

$$\left.\begin{aligned}M_L &= \sum_{i=1}^{N} C_{iL}M_i\\ M_g &= \sum_{i=1}^{N} C_{ig}M_i\end{aligned}\right\} \tag{10-47}$$

where M_i denotes molecular weight of the ith component.

The molar flux density of the ith component is obtained by observing that any component may be transported in both liquid and gas phases. Thus

$$\hat{v}_i = C_{iL}\,\frac{\rho_L}{M_L}\,\hat{v}_L + C_{ig}\,\frac{\rho_g}{M_g}\,\hat{v}_g \tag{10-48}$$

is the flux of i in moles per unit area per unit time. Thus the continuity equation yields for every component ($i = 1, 2, \cdots, N$)

$$\nabla\cdot\left[C_{iL}\,\frac{\rho_L}{M_L}\,\hat{v}_L + C_{ig}\,\frac{\rho_g}{M_g}\,\hat{v}_g\right] = -\phi\,\frac{\partial}{\partial t}\left[C_{iL}\,\frac{\rho_L}{M_L}\,S_L + C_{ig}\,\frac{\rho_g}{M_g}\,S_g\right] \tag{10-49}$$

where

$$S_L + S_g = 1 \tag{10-50}$$

if no connate water is included.

Now from the discussion of section 10.30 it is evident that in petroleum reservoirs we can safely assume the liquid and gas in any tiny region of the reservoir to be in phase equilibrium. Thus

$$C_{ig} = E_i C_{iL} \tag{10-51}$$

where the equilibrium ratios are functions of the composition in this tiny region of the reservoir, i.e. at a given point.

Here also it must be decided whether the equilibrium ratios corresponding to differential or flash vaporization are to be used. The proper values are probably well approximated by the differential process values. Also we must recognize the equilibrium ratios, as ordinarily obtained, correspond to vapor and liquid having the same pressures. Here, due to interfacial tension, p_L and p_g are not equal. This could probably be corrected for in a manner similar to that discussed in the treatment of vapor pressure in section 2.40. On the other hand we can with some unknown error ignore capillary effects and assume p_L and p_g equal.

Now equations (10-14) and (10-17) can be used to express C_{iL} and C_{ig} in terms of total composition, thus

$$C_{iL} = \frac{C_i}{1 + C_g(E_i - 1)}$$

and

$$C_{ig} = \frac{C_i E_i}{1 + C_g(E_i - 1)}$$

in our present notation. Here C_g is mole fraction of total system in the gas phase and C_i is mole fraction of total system which is component i, both for a tiny local region (a point) in the reservoir.

We have the subsidiary conditions

$$\left. \begin{array}{l} \displaystyle\sum_{i=1}^{N} C_i = 1 \\[2mm] \displaystyle\sum_{i=1}^{N} C_{iL} = 1 \\[2mm] \displaystyle\sum_{i=1}^{N} C_{ig} = 1 \end{array} \right\} \tag{10-52}$$

and

$$C_L + C_g = 1 \tag{10-53}$$

It is important to recognize that the over-all composition within a tiny region (a pore), is not the same throughout the reservoir.

If capillary pressure is neglected then the following system of equations is obtained

$$\nabla \cdot \left\{ \left(\frac{\frac{\rho_L K_L}{\mu_L M_L} + \frac{\rho_g K_g}{\mu_g M_g} E_i}{1 + C_g(E_i - 1)} \right) C_i \nabla p \right\} = \phi \frac{\partial}{\partial t} \left\{ \left(\frac{\frac{\rho_L S_L}{M_L} + \frac{\rho_g S_g E_i}{M_g}}{1 + C_g(E_i - 1)} \right) C_i \right\} \quad (i = 1, 2, \cdots N) \quad (10\text{-}54)$$

with

$$\sum_{i=1}^{N} C_i = 1$$

$$S_g + S_L = 1$$

and also, quite obviously,

$$C_g = \frac{\frac{\rho_g}{M_g} S_g}{\frac{\rho_g}{M_g} S_g + \frac{\rho_L}{M_L} S_L}$$

The E_i, ρ_g, M_g, ρ_L, M_L, μ_g and μ_L are to be treated as functions of the C_i, p and T. For isothermal conditions the $N + 3$ quantities, C_i, $i = 1, 2, \cdots N$, p, S_L and S_g are to be determined as functions of the space coordinates and time. To this end we have N equations of the form (10-54) plus the two equations following (10-54). This totals only $N + 2$ equations which is one too few. The additional equation is provided by

$$\sum_{i=1}^{N} C_{iL} = 1$$

or in terms of the C_i

$$\sum_{i=1}^{N} \frac{C_i}{1 + C_g(E_i - 1)} = 1$$

With this last equation a mathematically determinant system of equations, $N + 3$ equations and $N + 3$ unknowns, is obtained.

With appropriate boundary conditions this system of equations can be solved by numerical methods on a large high-speed digital computer. Until the present time no such solutions have been published in the literature. For a large number of components, N, this is a formidable problem. It is a certainty that these equations will be solved at some time in the near future because of the understanding of recovery processes which could result.

10.50: Second Formulation of the Solution-Gas Drive Process

The formulation of the solution-gas drive process considers the spatial distribution of each component as a function of time. An alternate treatment is possible in which this is not the case. This is constructed as follows.

In section 10.20 the distribution of a multicomponent system between the vapor and liquid phases was considered from the standpoint of gas solubility in oil. If we call all residual liquid at atmospheric conditions, as obtained by differential vaporization, oil, then we may call gas just gas without regard to composition. Then we investigate the spatial variation of gas and liquid saturation with time.

At any instant an element of the reservoir will contain a certain volume of liquid and a certain volume of gas. This liquid, if reduced to atmospheric conditions, would yield a volume of oil and a volume of gas. The ratio of this gas volume to oil volume is just the gas solubility, s.

If gas saturation is S_g and the liquid saturation is S_L, then in a unit volume of reservoir there is a mass of oil given by, $\phi \rho_{os} S_L / \beta_o$ where ρ_{os} is the oil density (at atmospheric condition, of course). In this same unit volume of reservoir there is a mass $\phi \rho_g S_g$ of free gas, where ρ_g is the mass density of free gas at p, T. Also there is a mass of gas dissolved in the oil; this mass is

$$\frac{\phi s \rho_{gs} S_L}{\beta_o}$$

Here ρ_{gs} is the mass density of the gas which would be evolved by reducing the liquid to standard atmospheric conditions. This density is measured at atmospheric conditions. Thus, the mass of gas per unit volume of reservoir is given by:

$$\phi \rho_g S_g + \frac{\phi s \rho_{gs} S_L}{\beta_o}$$

Now ρ_{os}, ρ_g, ρ_{gs}, s and β_o are all functions of over-all composition and, as was pointed out in section 10.40, this over-all composition is not uniform throughout the reservoir. Also in any element of the reservoir the composition will be changing with time. If we ignore this variation in over-all composition with position and time, we can obtain a much simpler formulation of the solution-gas drive process than that given in the previous section.

Thus by considering the two "components," oil and gas, we have

$$\phi \rho_g S_g + \frac{\phi s \rho_{gs} S_L}{\beta_o}$$

for the mass of "gas" per unit volume of reservoir, and

$$\frac{\phi \rho_{os} S_L}{\beta_o}$$

for the mass of "oil" per unit volume of reservoir.

In the reservoir there are two flowing phases. Oil is transported only in the liquid phase while gas is transported in both liquid and gas phases. Thus the mass flux density of oil is by Darcy's law

$$-\frac{\rho_{os}}{\beta_o}\frac{K_L}{\mu_L}(\nabla p_L + \hat{1}_v\rho_L g)$$

Here ρ_L is the mass density of the liquid which is a function of pressure and temperature since a unique over-all composition is assumed fixed.

Similarly, the mass flux density of gas is

$$-\rho_g\frac{K_g}{\mu_g}(\nabla p_g + \hat{1}_v\rho_g g) - s\frac{\rho_{gs}}{\beta_o}\frac{K_L}{\mu_L}(\nabla p_L + \hat{1}_v\rho_L g)$$

Both here and above μ_g and μ_L are functions of pressure and temperature because a fixed uniform over-all composition is assumed.

Here as in the previous formulation of this process a problem arises with the assumption of different pressures in liquid and gas phases. Thus solubility, formation volume factor, etc., are determined experimentally with both phases under the same pressure. Either capillary pressure must be ignored or else a correction must be worked out for capillary effects on solubility, etc. Here we will ignore capillary pressure and put $p_g = p_L = p$.

We will also ignore gravity effects and hence put the acceleration of gravity, g, equal to zero in the flux terms above.

Now the conservation of mass of oil yields in the continuity equation

$$\nabla \cdot \left(\frac{\rho_{os}K_L}{\beta_o\mu_L}\nabla p\right) = \phi\frac{\partial}{\partial t}\left(\frac{\rho_{os}}{\beta_o}S_L\right) \tag{10-55}$$

and for conservation of mass of gas

$$\nabla \cdot \left[\left(\frac{\rho_g K_g}{\mu_g} + \frac{s\rho_{gs}K_L}{\beta_o\mu_L}\right)\nabla p\right] = \phi\frac{\partial}{\partial t}\left(\rho_g S_g + \frac{s\rho_{gs}S_L}{\beta_o}\right) \tag{10-56}$$

These two equations in conjunction with $S_L + S_g = 1$ constitute three equations for determining the three quantities, p, S_L and S_g as functions of the space coordinates and time.

It is important to remember that these equations are approximate since we are considering the over-all composition to be fixed and uniform. That is, gas is "gas" and oil is "oil" without regard to composition. Thus the same functions of p and T: s, β_o, etc., apply at all points and all times and this is not strictly correct.

Before these equations can be solved it is necessary to specify initial and boundary conditions. These we deduce from physical considerations. Since first-time derivatives of both p and S occur, it is necessary to give

$$p(x,y,z,0) = p_o(x,y,z) \tag{10-57}$$

and

$$S_L(x,y,z,0) = S_{Lo}(x,y,z) \qquad (10\text{-}58)$$

where p_o and S_{Lo} are the initial distributions of p and S_L.

Also, since these differential equations represent the conservation of "oil" and "gas" respectively, it is necessary to specify

$$\left(-\frac{\rho_{os}K_L}{\beta_o\mu_L}\nabla p\right)\cdot\hat{n} = (\text{mass flux of oil normal to boundary}) \qquad (10\text{-}59)$$

and

$$\left[-\left(\frac{\rho_g K_g}{\mu_g}\frac{s\rho_{gs}K_L}{\beta_o\mu_L}\right)\nabla p\right]\cdot\hat{n} = (\text{mass flux of gas normal to boundary}) \qquad (10\text{-}60)$$

on the boundaries.

Numerical solutions of this problem have been computed by Sheldon, Garvin and West,[7] and also by Stone and Garder[6]. The nature of their results is discussed in the next section. In this connection it should be noted that the specification of p on a boundary is also a possible boundary condition.

10.60: Third Formulation of the Solution-Gas Drive Process; The Material Balance Equations

It has been shown that if we call gas, "gas" and oil, "oil", both without regard to composition, and also assume unique functions, $s(p, T)$, $\beta_o(p, T)$, $\mu_L(p, T)$, etc., then two partial differential equations are obtained which describe the pressure and fluid saturations as functions of position and time, i.e. equations (10-55) and (10-56).

If each of these is integrated over the volume of the reservoir a very useful result is obtained. The volume of the reservoir is bounded by two types of boundaries, those across which flow occurs (well bore surface, for example) and those across which no flow occurs (impermeable boundaries). Consider the volume integral of equation (10-55)

$$\int_{V_R}\nabla\cdot\left(\frac{\rho_{os}K_L}{\beta_o\mu_L}\nabla p\right)dV_R = \phi\int_{V_R}\frac{\partial}{\partial t}\left(\frac{\rho_{os}S_L}{\beta_o}\right)dV_R \qquad (10\text{-}61)$$

where V_R denotes reservoir volume. The volume integral on the left can be transformed by the divergence theorem into a surface integral over the surfaces bounding V_R. There results

$$\int_{A_R}\frac{\rho_{os}K_L}{\beta_o\mu_L}\frac{\partial p}{\partial n}dA_A = \phi\int_{V_R}\frac{\partial}{\partial t}\left(\frac{\rho_{os}S_L}{\beta_o}\right)dV_R \qquad (10\text{-}62)$$

where $\partial p / \partial n$ is the directional derivative of p normal to the surface A_R which bounds V_R, i.e. the component of ∇p normal to the bounding surface. This normal direction is outward from the volume V_R.

If none of the bounding surface A_R is in motion (A water-oil interface would be in motion during natural-water drive, for example) then the time derivative on the right can be brought outside the integral.* Then if the boundary has only two parts, one of area A across which flow occurs and another across which no flow occurs, we have

$$\int_A \frac{\rho_{os} K_L}{\beta_o \mu_L} \frac{\partial p}{\partial n} \, dA = \frac{\partial}{\partial t} \int_{V_R} \phi \frac{\rho_{os} S_L}{\beta_o} \, dV_R \tag{10-63}$$

The quantity on the left is the total flow rate of oil mass into the reservoir and the quantity on the right is the rate of change of total oil mass in the reservoir.

If we replace all quantities by averages, averages over the area A on the left, and averages over the volume V_R on the right, we obtain

$$\left[-\frac{\rho_{os} K_L}{\beta_o \mu_L} \frac{\partial p}{\partial n} A \right]_A = -\frac{\partial}{\partial t} \left[\phi \frac{\rho_{os} S_L V_R}{\beta_o} \right]_{V_R} \tag{10-64}$$

Here the subscripts indicate the regions over which the averages are taken.

In a similar manner we obtain from equation (10–56)

$$\left[-\left(\frac{\rho_g K_g}{\mu_g} + \frac{s \rho_{gs} K_L}{\beta_o \mu_L} \right) \frac{\partial p}{\partial n} A \right]_A = -\frac{\partial}{\partial t} \left[\left(\rho_g S_g + \frac{s \rho_{gs} S_L}{\beta_o} \right) \phi V_R \right]_{V_R} \tag{10-65}$$

The left members of these two equations represent the mass rates of flow from the reservoir of oil and gas, respectively. The right members are the rates of decrease of mass in the reservoir of oil and gas, respectively. These equations are one form of the *material balance equations*. Other forms are readily obtained simply by including oil-water boundaries (an aquifer) for water drive and gas-oil boundaries (for gas-cap drive). These equations are discussed in considerable detail in many texts on petroleum reservoir engineering.

Now we will consider the solution of these material balance equations and their relation to the more exact treatment of the previous section. First of all it is obvious that if saturation and pressure are essentially uniform throughout the reservoir then the approximation on the right in these equations is essentially exact.

The actual manner of solution of these equations is most easily seen by

---

* Strictly speaking, equation (10-62) can be derived directly in which case it is seen that $\partial / \partial t$ should be outside the integral in all cases and can be taken inside only for fixed boundaries.

writing them in the form

$$Q_o = -\left(\frac{\phi V_R}{\beta_o}\right)\frac{dS_L}{dt} - \left[\frac{V_R S_L \phi}{\rho_{os}}\frac{d}{dp}\left(\frac{\rho_{os}}{\beta_o}\right)\right]\frac{dp}{dt} \tag{10-66}$$

and

$$R_{go}(p, S_L)\,Q_o = -\frac{\phi V_R}{\rho_{gs}}\left(\frac{s\rho_{gs}}{\beta_o} - \rho_g\right)\frac{dS_L}{dt} - \frac{\phi V_R}{\rho_{gs}}\left[\frac{d\rho_g}{dp} + \frac{d}{dp}\left(\frac{s\rho_{gs}}{\beta_o} - \rho_g\right)S_L\right]\frac{dp}{dt} \tag{10-67}$$

where

$$R_{go}(p, S_L) = \frac{\dfrac{\rho_g K_g}{\rho_{gs}\mu_g} + \dfrac{sK_L}{\beta_o\mu_L}}{\dfrac{K_L}{\beta_o\mu_L}} \tag{10-68}$$

is the gas-oil ratio at pressure, p (volume ratio), and Q_o is volume flow rate of oil at p.

Thus, for example, if Q_o is specified as a function of t, and p and S_L are given at $t = 0$ these equations can be solved numerically.

Typical solutions of this type are shown in Figures 10.6 and 10.7. Shown for comparison are the numerical solutions of the partial differential equa-

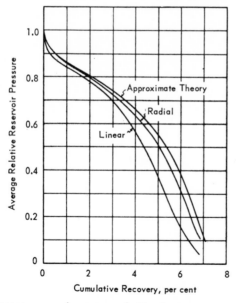

Figure 10–6. Average reservoir pressure (ratio of average pressure to initial pressure) versus cumulative recovery for solution-gas drive reservoir. Comparison of linear and radial reservoirs to the material balance calculation. (*After West et al., 1954.*)

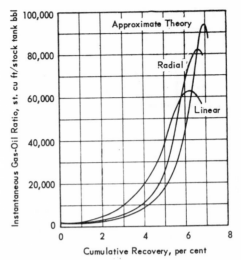

Figure 10–7. Gas-oil ratio versus cumulative recovery for solution-gas drive reservoir. Comparison of linear and radial reservoirs to the material balance calculation.

tions described in Section 10.50. These results were published by Sheldon, Garvin and West[6] as a comparison between the approximate method described here and the more exact theory as described in Section 10.50.

The two cases of numerical solution of the partial differential equation illustrated here are a linear reservoir and a radial reservoir with the same fluids and initial and boundary conditions as employed for the approximate, or material balance, calculation. The general character of the results for all three systems is quite similar. Stone and Garder[6] have made a detailed study of the numerical solution of the partial differential equations of solution-gas drive. In particular they included gravity in their calculations where West, Garvin and Sheldon did not. Also Stone and Garder compared calculated results to results obtained with a laboratory model and found that, in certain cases, supersaturation is not as severe as expected.

For a more detailed discussion of the solution of the partial differential equations the reader should consult the cited papers. For a complete discussion of material balance methods Muskat's text[4] should be consulted.

10.70: Flow with Change of Phase; General Procedures

The treatments of the solution-gas drive process presented in previous sections of this chapter serve to illustrate the basic physical considerations involved in the formulation of flow processes including phase transitions. In a very similar fashion problems involving non-equilibrium phase relations or chemical reactions can be treated. Basically all that is required is a knowledge of the physical reactions or transitions as occur in stationary systems. Then these reaction laws are combined with Darcy's law of flow and appropriate conservation equations, i.e. continuity equations.

EXERCISES

1. Show that since the equilibrium ratios of a hydrocarbon system are particular functions of p and T a scaled model study of solution-gas drive should be conducted with actual reservoir fluids and at reservoir temperature and pressure.
2. Show that a scaled model study of solution-gas drive should be conducted at flow rates comparable to reservoir rates.

References

1. Burcik, E. J., "Properties of Petroleum Res. Fluids," John Wiley & Sons, N.Y. (1957).
2. Glasstone, S., "Textbook of Physical Chemistry," D. Van Nostrand Co., New York (1946).
3. Higgins, R. V., *Trans. AIME*, **201**, 231 (1954).
4. Muskat, M., "Physical Principles of Oil Production," McGraw-Hill Book Co., N.Y. (1949)..
5. Pomeroy, R. D., Lacey, W. N., Scudder, N. F., and Stapp, F. P., *Ind. Eng. Chem.*, 1015, Sept. 1933.
6. Stone, H. L., and Garder, A. O., Jr., Paper No. 1518-G, presented at Fall Meeting AIME, Denver, Colo., Oct. 1960. To be published in *Trans. AIME*.
7. West, W. J., Garvin, W. W., and Sheldon, J. W., *Trans. AIME*, **201**, 217 (1954).

INDEX